William John Loftie

Inigo Jones and Wren

Or, the Rise and Decline of Modern Architecture in England

William John Loftie

Inigo Jones and Wren
Or, the Rise and Decline of Modern Architecture in England

ISBN/EAN: 9783337087630

Printed in Europe, USA, Canada, Australia, Japan

Cover: Foto ©berggeist007 / pixelio.de

More available books at **www.hansebooks.com**

INIGO JONES AND WREN

OR

THE RISE AND DECLINE
OF
MODERN ARCHITECTURE IN ENGLAND

STAIRCASE, ASHBURNHAM HOUSE. BY INIGO JONES.

INIGO JONES AND WREN

OR

THE RISE AND DECLINE

OF

MODERN ARCHITECTURE IN ENGLAND

BY

W. J. LOFTIE

AUTHOR OF 'A HISTORY OF LONDON,' ETC.

London

RIVINGTON, PERCIVAL & CO.

1893

PREFACE

It is, perhaps, necessary to explain why the term "Palladian" is here chiefly used for the kind of architecture practised by Inigo Jones and Wren, to which the following pages relate. The only other possible word, "renascence," or "rennaissance," is not sufficiently definite, and has moreover a foreign sound. Some people speak of "Queen Anne," but the style was in vogue here in the time of Queen Anne's great-grandfather, nearly a century before her glorious reign. The art, as described by Palladio, and as practised by himself at Vicenza, and by his contemporaries and followers in Venice, Padua, Genoa, Florence, Rome, and many other places, is easily recognised. Its influence was nowhere more marked than in England, which indeed may be termed its second home, and where it flourished even better than in its birthplace. The French, though their writers refer to the fact, seem never to have cared for it as we have done; and the work of their great architects, Mansard, Perrault, Le Mercier, and their fellows, differs in many important particulars, nay, in fundamental principles, from that of Inigo Jones, Wren, Burlington, or Chambers, whose art was essentially English, though founded on Italian. As the word "Palladian" then conveys a definite idea and is moreover easy to pronounce, and as after much seeking I have found no other name so suitable, I hope I may be excused for using it. "Queen Anne" has a limited, "Italian" an unlimited meaning. True, the faults of Palladio, as set forth, not so much in his drawings as in his actual buildings, are easily discovered, though they hardly concern us here. He was careless of details, his ornaments are often coarsely cut, and though he had laid down such exact rules for proportion he was always ready to break them himself when occasion arose. It is perhaps this freedom as much as anything which recommended his views on the Italian style, itself founded on the Roman and that on the Grecian, to the admiring notice of posterity. There can be but little doubt that the publication of his book in 1570 led to its adoption here, as representing learned or classical art as distinguished from Gothic.

Another excuse or apology I have also to make. This book is not written for architects, nor is it by an architect. My earnest hope in launching it upon the world is that it may reach some of those by whom architects are employed. I do not doubt my critics, if I have any, will object that I have not used the correct terms in describing the architectural features of some buildings. But I have been advised that such terms are often

Preface

only a puzzle to the general reader; that, as he is not about to design, they are superfluous, and that much talk about friezes, triglyphs, drops, modillions, architraves, entablatures, and so on, would only be a weariness and interrupt the course of the narrative or argument. At the same time, in several cases, I have added a more complete and technical account of any object which seemed to require it.

I should conclude this preface by saying that as I was writing the last lines of my last chapter, the book on the question, *Is Architecture a Profession or an Art?* was put into my hands. I feel obliged to agree with nearly every word of it, and especially with Mr. Norman Shaw's essay. I am glad to find so many architects ready to recognise the place and importance of art in design; and though the volume came out too late to be of any advantage to me, I feel it is a cause of deep satisfaction that so powerful a movement should have been made, and by such eminent artists, to lay down distinctly the very principles on which every chapter of this book has been written. I cannot but think that before long the employers of architects will be brought to see that beauty in design is better than ornament, and far less costly.

I am not an architect, as I have said, but a member of the general public, though I belong to a profession the members of which as a class give the most employment to architects. I ought perhaps to apologise for venturing, even after many years of study, to address the public on the subject; but however eminent an architect may be, it is not the architect but his employer, the amateur, who is entitled to make the final decision; and it cannot be doubted that much if not all of the bad building of the day is due to the ignorance or indifference of the architect's employers.

I have endeavoured, I hope with success, to unravel the history of Inigo Jones's two great designs for Whitehall, which have so completely puzzled previous writers, and I have endeavoured also to apply similar principles to the elucidation of the different schemes made by Wren for St. Paul's.

The illustrations are mainly from the plates published during the golden age of English Palladian. They are, however, largely supplemented by photographs, especially of those charming buildings of the transitional period which are to be found in the west country, and where the Bath stone forms such a ready vehicle for the expression of poetry in stone. Mr. R. Wilkinson, of Trowbridge, has obligingly placed the results of many years' photography at my disposal, and I beg to thank him warmly. The London Stereoscopic Company have also permitted me to have copies made from their productions, and I beg very gratefully to acknowledge their courtesy. The two prints from the works of Inigo Jones, which were published by the Society for Photographing Remains of Old London, I owe to the kindness of Mr. Alfred Marks.

CONTENTS

I

INTRODUCTION

Modern Gothic—Windsor Castle—Palace of Westminster—New Law Courts Detail and Proportion—Salisbury and Chichester—The City of London—Ornament—Architectural Teaching—A Contrast—Churches—Gothic Churches—Wren's Churches—Novelty—The Anomalous or Eclectic Style—The chief want of Modern Architecture *Page 3*

II

THE DECAY OF GOTHIC

"An arch never sleeps"—Progress of the Pointed Arch—Objects of Gothic Builders—Flat Arches—Rules—Modern Imitations—The Albert Memorial—The last Gothic—The so-called "Debased Style"—Late Gothic at Oxford—The Staircase at Christ Church—The Tom Tower—Late Gothic at Cambridge—Bath Abbey—Hospital at Corsham—Charles Church, Plymouth—Hampton Court—"Peter Torrysany"—St. James's Palace—Middle Temple Hall—Inigo Jones—Wren . . 19

III

ELIZABETHAN ARCHITECTURE

A Time of Change—A New Style—Elizabethan Houses—Examples—The Irregular Type—The Regular Type—Haddon Hall and Longleat—The Duke's House, Bradford—Cheshire Houses—John of Padua—The Masons—Lord Burghley.

Page 51

IV

THE BEGINNINGS OF PALLADIAN

The Beginnings of Palladian—The First Examples—Tombs by Torregiano—Sir Anthony Browne's Monument—Mantelpieces—The Royal Exchange—Caius College—Recent Vandalisms—The Gates of Humility, Virtue and Knowledge, and Honour—Palladian in Fashion—John Shute—Lomazzo—Birth of Inigo Jones—Gothic and Palladian—Palladio—Vitruvius—Proportions of the principal Orders

79

V

INIGO JONES

A List—Parentage and Name—Birth and Baptism—Visits Italy—A Landscape Painter—Proportion—In Denmark—With Prince Henry—A Scene Painter—Surveyor-General—Numerous Drawings—Method of Working—Stage Experience—Arch Row, Lincoln's Inn Fields—Greenwich—Somerset House—York House—Jones and Stone—New Palace of Whitehall—Design for James I.—Design for Charles I.—The Banqueting Hall—A Reredos—Old St. Paul's—St. Paul's, Covent Garden—Ashburnham House—Country Houses—School of Inigo Jones—Death and Burial

109

VI

WREN

Wren and Oliver Cromwell—Wren and Webb—The Chapel of Pembroke College, Cambridge—The Sheldonian Theatre—The Library, Trinity College, Cambridge—Wren at Paris—The Great Fire—Windsor—Chelsea—Greenwich—The Monument—Hampton Court—Kensington

151

VII

WREN'S CHURCHES

Obnoxious to Bishops—Many destroyed—Method of procedure—Case of St. Antholin's—A Monstrous Falsehood—Classification—St. Paul's—Court Influence—A Protestant Design—An Artificial Design—Decorations—Parish Churches—Two principal Patterns—Domed Churches—Gothic Churches.

Page 177

VIII

THE SUCCESSORS OF WREN

Vanbrugh—Hawksmoor—Gibbs—James—Archer—Burlington—Campbell—Kent—Taylor—Chambers—Adam—Wood of Bath—Baldwin—Palladian in the Provinces—Dublin—The Bank—The Four Courts—The Custom House—Trinity College—Barry's Club Houses—The Grecian Style—The Reign of Stucco—The New Gothic—Conclusion

215

ILLUSTRATIONS

	PAGE
Staircase, Ashburnham House. By Inigo Jones. From a Photograph by the Society for Photographing Remains of Old London . . *Frontispiece*	
Church of St. Mary-le-Bow, by Wren, with the Norman Crypt. From the Print published by the Society of Antiquaries . . .	2
South Wraxall. From a Photograph by Mr. R. Wilkinson .	25
Saloon, South Wraxall. From a Photograph by Mr. Wilkinson	29
Almshouse, Corsham, Wilts. From a Photograph by Mr. Wilkinson	37
Charles Church, Plymouth. From a Photograph .	41
Gallery, Haddon Hall. From a Photograph by the Stereoscopic Company .	57
Jaggard's Manor-House, Corsham, Wilts. From a Photograph by Mr. Wilkinson	61
The Duke's House, Bradford-on-Avon. From a Photograph by Mr. Wilkinson	65
Longleat. From a Photograph by Mr. Wilkinson .	69
Stewart Monuments, Ely Cathedral. From a Photograph .	81
South Wraxall. From a Photograph by Mr. Wilkinson	85
Gate of Honour, Caius College. From a Photograph	89
Thieni Palace, Vicenza. From Ware's *Palladio*	100
Almerico Palace, Vicenza. From Ware's *Palladio* .	101
Mocenigo Palace. From Ware's *Palladio* .	102
Proportions of Ionic, Corinthian, and Composite Columns. From Ware's *Palladio*	103
Palace of Whitehall. As designed by Inigo Jones, 1619. From Muller's Print	111
Lincoln's Inn Fields. From a Photograph by the Society for Photographing Remains of Old London	117
Part of Court, showing Banqueting Hall, Whitehall. From Kent's *Inigo Jones*	123

	PAGE
Portion of Design, Whitehall. From Kent's *Inigo Jones*	125
Portico, Old St. Paul's. From Kent's *Inigo Jones*	129
Covent Garden, Church and Piazza. From *Vitruvius Britannicus*, vol. ii.	133
Cobham Hall, Kent. From a Photograph	137
Brympton. From a Photograph by Mr. Wilkinson	141
Coleshill. From *Vitruvius Britannicus*, vol. v.	145
Greenwich Hospital. From a Photograph	161
Part of Wren's First Design for Greenwich. From *Vitruvius Britannicus*, vol. i.	165
Greenwich: Vanbrugh's Work. From a Photograph	181
The West Prospect of St. Paul's Church. From *Vitruvius Britannicus*, vol. i.	185
St. Paul's Cathedral: Wren's First Design	189
St. Lawrence Jewry. From a Photograph by the Stereoscopic Company	205
East Front of Blenheim. From *Vitruvius Britannicus*, vol. i.	217
Public Buildings at Cambridge. By James Gibbs	223
St. Mary-le-Strand. By Gibbs	225
Spencer House, Green Park. By Vardy. From *Vitruvius Britannicus*, vol. iv.	231
House by Lord Burlington for General Wade. From *Vitruvius Britannicus*, vol. iii.	235
Dormitory, Westminster School. By Burlington. From Kent's *Inigo Jones*	239
Assembly Rooms, York. By Burlington. From *Vitruvius Britannicus*, vol. iv.	243
Burlington House. From *Vitruvius Britannicus*, vol. iii.	247
Gate, Burlington House. From *Vitruvius Britannicus*, vol. iii.	251
Wrotham, Middlesex. By Isaac Ware. From *Vitruvius Britannicus*, vol. v.	255
Villa, Chiswick. By Lord Burlington. From Kent's *Inigo Jones*	259
Section, Chiswick. By Burlington. From Kent's *Inigo Jones*	263
Holkham, Norfolk. By Kent. From *Vitruvius Britannicus*, vol. v.	266
South Front, Kedleston. By Adam. From *Vitruvius Britannicus*, vol. iv.	269
Kedleston. By Adam. From *Vitruvius Britannicus*, vol. iv.	269
Prior Park, Bath. From a Photograph by Mr. Wilkinson	272
Reform and Carlton Clubs, Pall Mall. From a Photograph by the Stereoscopic Company	275
Church at Glasgow. By Thomson. From a Photograph	278

I
INTRODUCTION

CHURCH OF ST. MARY LE BOW, BY WREN, WITH THE NORMAN CRYPT.

I

INTRODUCTION

Modern Gothic—Windsor Castle—Palace of Westminster—New Law Courts Detail and Proportion—Salisbury and Chichester—The City of London—Ornament—Architectural Teaching—A Contrast—Churches—Gothic Churches—Wren's Churches—Novelty—The Anomalous or Eclectic Style—The chief want of Modern Architecture.

It may be assumed without much proof that the modern attempt to revive Gothic architecture has been a failure. Unfortunately, the influence of the movement has told also on other styles. Up to the end of the sixteenth century, while the old Gothic was still alive it was progressive. It never rested. It was always seeking and often finding improvement. The architects of the first buildings in the pointed style bequeathed their traditions to their pupils, and the chapel of Henry VII. is lineally descended from Salisbury Cathedral. In the modern Gothic there was no such tradition. An architect sat down to design. He did not say, "I will try if I can improve on my last work, or, on the work of my predecessor." Quite the contrary. He said, "I will design this building in the style of the thirteenth century. I will design that one in the style of the fourteenth. This church shall be Decorated. That church shall be Perpendicular." All this was essentially false in art. We should laugh at a painter or a composer who went to work in this way. Gothic architecture, nevertheless, found favour with many people of taste, and was eagerly taken up by architects. The reasons are easily found. The great patrons of art hoped to obtain buildings like those of the past. They thought architects who tried to imitate mediaeval buildings would be able to build as Bishop Poore, or as William of Wykeham built. But the architects constantly disappointed their employers. The revival has not produced one really good or beautiful building. The best, strange to say, is Windsor Castle, which is universally abused as bad Gothic. It may be so, indeed it would be hard to say it is not, but Windsor Castle is essentially picturesque, and in this it stands alone among the efforts of the modern Goth. Viewed from a distance, it is a fair form in a fair landscape. Viewed near at hand, it certainly has its faults of detail. The windows of Salvin, and the so-called "restorations" of Scott have not improved it; but from the gateway of Henry VIII. to the uttermost verge of the east front, it is a long series of picturesque and scenic arrangements, one feature enhancing another, and affecting the mind of the visitor as no other modern building can, whether Gothic, Grecian, or Palladian.

It is necessary here, however painful, to notice some other results of the Gothic revival. Of these, the most successful is the Palace of Westminster. No doubt it is marred by many grievous faults, but it affords us a clue to the general failure of this new Gothic style — a style I have elsewhere called Vandalic, to distinguish it from the original pointed style, with which, however, I have no special concern just yet. The Palace of Westminster is successful in two respects. Wherever Sir Charles Barry gave way to the Palladian tradition, in which he had been brought up and lived till then, he was successful. Wherever, on the other hand, he gave way to the new Gothic teaching, he failed. This can be proved in a moment. The river front is allowed by all to be the worst feature of the whole design. Here his Palladian instincts would have prescribed an imposing central mass with wings, not of necessity absolutely symmetrical. Instead, as the eye travels toward the centre, the building becomes, or seems to become, which is, of course, precisely the same thing, lower and lower, meaner and meaner; and the fine proportions of the end towers are abandoned and lost. The successful parts of the building are those in which proportion has been most carefully calculated. The square tower, with its grand double archway, and, in the interior, the lobbies of both houses, and the houses themselves, have proportions of the simplest and best kind. The lobbies are cubes, and the houses are, or were, —for the House of Commons has been altered,—double cubes.

Another great Gothic building is a still greater failure. The one fine external feature of the new Law Courts is the lofty gable of the central hall; but it is so masked and interfered with by the crowd of minor buildings which surround it and block it up, that it can hardly be distinguished. The interior of the hall, though too gloomy, is unquestionably very fine, but here, too, the absence of proportion makes it look smaller than it is, and robs it of much of its effect.

The teachers of the new Gothic said, in short, that proportion could take care of itself—that detail was everything. This is the doctrine of Mr. Ruskin. It was the doctrine of Sir Gilbert Scott and of Augustus Pugin. As a consequence, we have such buildings as the St. Pancras Hotel and the

Albert Memorial covered with ornament, but far indeed from being themselves ornamental. The heraldic shields on the river front of the Houses of Parliament, alone must have cost as much in thought, design, skill, and money as would have made the front ornamental, without so much as a moulding. But this attention to detail, this substitution of shadow for reality, was characteristic of the work of the whole school. Mr. Ruskin, for example, entitles his great architectural essay *The Stones of Venice*, and the name exactly describes the book. The bitterest scoffer could not have done it better. The teaching is to think of the stones, of the marble, the porphyry, the carving, the colour, the minute points of sentiment, and let everything else—building, proportion, mass, stability, lightness—take care of themselves. This was not the teaching of the real Gothic architects. Look at Salisbury Cathedral for an example, though it is a late work of the Early English School. Though for a hundred years past a long succession of architects of the modern so-called Gothic School, from Wyatt to Scott, have done their worst, they have not been able to rob the church of its admirable proportions. Nothing short of absolute destruction can do that. We have a case in point. The spire of Chichester Cathedral—a building almost contemporary with Salisbury, and in some respects, though on a smaller scale, so like it that a local proverb says, the master built Salisbury, the man built Chichester—fell down bodily. Scott was entrusted with the rebuilding. Every safeguard that could be devised was set up to prevent his altering any detail of the old steeple. But he eluded the vigilance of his employers, and changed the proportions with disastrous effect; nor was it possible to persuade him that a mere raising of the tower a few feet, a mere lowering of the spire, could in any way influence the result. The details, the tracery, the mouldings are all the same: these in Scott's mind, no doubt, were the essentials; but the tower of Chichester Cathedral, as the citizens fondly remembered it, is a thing of the past.

The pernicious influence of this teaching has spread itself far beyond the boundaries of the modern Gothic style. A walk through the City of London is a painful exercise to any architectural critic on this account. Millions have been lavished during the past few years on new buildings,—piles of offices, and banks for the most part,—nearly the whole of the City has been rebuilt, and money has evidently been no object in comparison with magnificence; and yet it may be safely asserted that not more than three buildings possess any quality which would leave them even tolerable if stripped of their ornamental details—polished granite, coloured marbles, gilded bronze.

The fault, it must be acknowledged, is not wholly that of the modern architect. It lies in the want of taste of the architect's employers. Wren or Inigo Jones would starve at the present day. Neither of them thought of ornament as an end in design. No *Stones of Venice* existed in their day to wax eloquent over the undercutting of a moulding or the meaning of a sculptured bracket. The architect who wants to be successful must please his public. If his public, like Swift's Yahoos, gloat on shining and coloured stones, why should he trouble himself to see that they have good classical proportions? There are three pairs of pillars side by side on the slope of a hill in Piccadilly. They are all of polished Syenite, all in the Tuscan Doric style, and each pair as we descend is a little longer than its predecessor. The effect is more than frightful. It is aggressively hideous, like the instruments of an ordinary German band all tuned in different keys. Even if any one pair of the pillars,—so elastic is this Tuscan Doric, Wren's favourite style, for a reason I shall endeavour to state further on,— might look well by itself, the three pairs, "tuned," so to speak. "in different keys," set the teeth on edge. The pursuit of ornament for its own sake is not therefore to be wholly attributed to the architects of the day. True, the teachers of the young architect are able to do nothing for the education of his taste. They were themselves brought up on the principles of the modern Gothic style. They have no chance of learning, still less of teaching, anything better. A short while ago I made a careful inspection of the drawings submitted by the younger generation of architects for an important prize. There was one Palladian design—it was for a town house—in which, though it was far from perfect, the student had drawn his inspiration from Burlington or Wren. This drawing, which showed no ornaments, was not even mentioned by the judges. The prize went to a design which was wholly devoid of proportion, wholly devoid of "style," wholly commonplace and vulgar; but the competitor had plastered his work all over with friezes, with mouldings, with wreaths, busts, jars, urns, and especially with bosses of coloured and polished granite and marble. He hit the taste of the day, the taste which has given us the new Natural History Museum, the Imperial Institute, St. Thomas's Hospital, and scores of other buildings in which ornament, the more anomalous the better, is predominant, style and proportion nowhere.

If by any means taste could be improved, even a little, not so much in the architects as in their employers, there would be some hope for the future of English architecture. The supply of good design would speedily rise to the demand. But the rich man in want of a house deliberately prefers what is bad. In the Bayswater Road there is at this moment an example which seems as if it had been specially constructed for my use here. I will not refuse it. About twenty years ago, an architect who understood his art built a house facing Kensington Gardens, and, being really an architect, not a mere ornamental builder, he made it what a moderately large town house ought to be—plain, solid, well-proportioned, but without any exterior ornament, or anything to attract and retain London soot. Driving past, one always looked out for it as an oasis in a desert of stucco and brown brick. There came a rich tradesman and built a house next door. It has carvings, wrought and cast iron railings, oriel windows, bulls' eyes, and a medley of architectural features of all kinds mixed up in a frantic but futile effort to attain the picturesque. One almost pities the architect. He has piled on everything he can think of—mullions, broken pediments, dormers, string-courses, cornices—and all in vain. One thing is lacking—proportion. The plain house next door eclipses it utterly. It is a costly eyesore, a thing to avoid, a house that may be comfortable enough to inhabit but is hideous to look at. Though loaded with ornament, it is not in the least ornamental—very much the reverse. When John Opie advised a young artist to mix his colours with brains, he hit the ordinary defect in all unsuccessful design, whether in pictures or in architecture. The Gothic revival injured architecture just for the reason that brains were not necessary in order to build as its votaries built. All the more difficult branches of the art were neglected. At first the public objected, and modern Gothic was not immediately popular. But by degrees it superseded every other style for churches, and even prevailed occasionally for private houses and public buildings. I feel great hesitation in stating my opinion that the style is unsuitable, except in one particular, for churches. It is said to be cheaper than any other, and as churches have but too often to be built where money is scarce, if this is true there is little more to be said. But a handsome Gothic church is as expensive as any other building can be, and a plain Palladian church like Wren's chapel recently ruined at Cambridge is very cheap as far as money cost is concerned, though it is expensive in thought and calculation and all those other necessaries which have become so scarce at the present day. In building Gothic churches, as for example St. Mary Abbots at Kensington, the architect has, in order to make his work look mediaeval, to surround it with chapels and side aisles and so forth; but he builds them in the same style as the central church itself, and so at once destroys any illusion he may have created, because in a genuine ancient Gothic church the chapels must of necessity be of a different date and therefore of a different style.

But apart altogether from any question as to Gothic examples, the more abstract question remains as to which is the best style for churches. For Roman Catholic worship the Gothic style is very suitable, though not certainly the best. Our beautiful old English parish churches were designed for the celebration of masses. It may, however, be remembered that many beautiful churches in Italy and elsewhere, were designed for the selfsame purpose in the various forms of the Palladian style. But for Protestant services one of Wren's patterns, such as St. Lawrence, is very preferable. Wren used the Gothic plan for some of his finest efforts; but even in them it is evident that he aimed at making a place suitable for reading and preaching, not a place suitable for the celebration of masses.

From this point of view, modern Gothic churches with mock side chapels are a failure. When we enter such a church, we see that the chapels only exist on the outside. They are "ornamental" features only. The whole interior forms one large chamber cut up into aisles by columns which only serve to intercept the view and interrupt the voice. Nothing, as I have said, seems to me so suitable for Protestant, and especially for Church of England worship, as one of Wren's city churches. His object always was to accommodate the largest congregation in such a way that all should be able to see and hear. In this respect the chapel of Pembroke College, Cambridge, came as near perfection in convenience as it was possible. When convenience could be combined with architectural beauty, as in St. Stephen's, Walbrook, the result was absolute perfection. Both these buildings have been senselessly altered of late years, and Wren's proportions lost, but we can still judge with an effort of what they formerly were, and wonder also that any one calling himself an architect could be found willing to lay his sacrilegious hands on them; but it is probable that in neither case did the architect, brought up in the modern school of architecture, know what he was doing.

This brings me to the last point I need touch upon in this preliminary chapter. The education of the modern architect teaches him anything except architecture. He may learn all about specifications. He sometimes knows beforehand what a building will cost, though even in this he often fails, as in a celebrated recent case. After the building, sanitary work, materials, and other mechanical parts of his design have been

settled, his functions cease. But some there are who aim at
novelty. It is this idea of seeking for originality which gives us
the worst things which deface our streets. A new style has
unquestionably been invented. It consists chiefly in a rebellion
against old authority. Its professors, I believe, term it "eclectic."
It might better be called the anomalous style. No rules of art
exist which it does not break. It has hitherto been accepted as
an axiom that an architect should make a chamber look larger
than it is, and masonry stronger than it is. Every effort should
be used to enhance size and give an idea of security. The new
school believes in none of these things. A solid building, they
think, should be made to look as flimsy as possible. Above all,
anything savouring of style must be avoided. The new Town
Hall at Manchester is an excellent example. By contrivances of
exceeding subtlety, the architect has managed to make it appear
so low that you instinctively bow your head lest you should hit
the roof. You read with surprise and incredulity that it is some
forty feet high or more. It is so cleverly designed that it does
not look seven. The art of sinking in architecture is further
illustrated by the new Natural History Museum. It looks as if
a breath would blow it away. Yet, for all I know to the contrary,
it is as solid, and, I fear, as durable as any other building in
London. Mr. Waterhouse usually designs in what I understand
he is pleased to call the Gothic style. Besides the Town Hall,
there are other buildings of his at Manchester, and some also by
pupils. The Manchester streets are oppressed by the heavy
mouldings, the shallow windows, the general air of having been
not built but cast, that pervades all their work. Mr. Collcutt, one
of the pupils, follows him after a fashion. If Mr. Waterhouse seeks
to pick out all that is least to be admired in Gothic, Mr. Collcutt

applies the same process to the so-called Italian style, and we have a new theatre in Shaftesbury Avenue, to say nothing of the Imperial Institute at South Kensington, in which the rare architectural features appear to be chosen not for their beauty, not for their fitness, but absolutely for their ugliness. The theatre occupies a place in a circus and its front is accordingly crescent shaped, the architect unfortunately forgetting that his building was to face the north. There is therefore no play of light and shade, and the front looks simply crooked, or as if the builder had been distracted. There is an immense display of absolutely unmeaning ornament culled from all sources. The windows are placed where they give a feeling of weakness, and there are oriels and turrets where they can produce no effect except to make the whole composition look as small as possible. The result is certainly original. There is not an axiom of the older architects that is not violated. The theatre looks cheap, and is dear. It looks low, and is tall. It is covered with ornament, and is ugly. Yet this is the "eclectic" style, and the architect has been commissioned to design one of the largest and most conspicuous edifices in the west of London.

Ignorance of the fundamental rules of proportion, as they were understood by Inigo Jones and Wren, is the great fault of the modern architect. His Gothic training has turned his attention to detail only. He thinks beautiful or well-imitated parts must produce a beautiful whole. He flounders about among small features and is surprised to find that the wildest combination will not result in picturesqueness, just as pressing down the whole key-board of a piano will not give harmony. Until proportion is again acknowledged and sought for as the most needful thing in architecture, and as something which may be reduced to mathematical rules like those which govern the sister art of music, we may have buildings, but we cannot have architecture. The Parthenon did not depend for its beauty on its sculptures. Its frieze and pediments are in the British Museum, yet it remains one of the most beautiful buildings of antiquity. Salisbury Cathedral has gained no picturesqueness by the restoration of its sculptured ornaments. St. Paul's would still be what it is in form and majesty if all Grinling Gibbons's flower wreaths were hacked away. But remove the carving and the moulded brick from a modern street front, and what would be left? It is a truism which cannot be too much insisted upon or too often repeated, that no amount of ornament will make an ugly building ornamental.

If the foregoing remarks are just, unconsciousness of proportion and harmony is the crying sin of the modern architect. If I could reach the ears of the patrons, the employers of the modern architect, I should not have protested in vain. But as long as the public taste is not offended by such horrors as the mock Gothic west front of Westminster Hall, or the mock Palladian of the National Liberal Club, or the utterly indescribable ugliness of the new School of Music, it is of little use to preach. When sound taste prevails, good design will be demanded, and, we may be sure, will be produced. It exists, but no one cares for it. True, an architect lately announced his brilliant discovery that some of our older public buildings owed almost all their charm to their delicate proportions, and that he was inclined to think that proportion would be better than ornament to beautify a building. So bold an innovator was, of course, pooh-poohed. His tone was considered almost

offensive to his brother architects, and his views, if acted upon,
would necessitate their learning mathematics and such like
rubbish. Nevertheless, there are signs that both the
public and their architects are beginning to see the necessity
of proportion, and also the possibility of taking up the
Gothic tradition where it was left by Wykeham and Bray.
Besides the new church in Sloane Street by the lamented
Mr. Sedding, which may be mentioned as one example
among several, we may note that lately a Soane medal was
given to a young architect for the design of a chapter house
on similar principles. Some architects also have plainly
recognised that the Palladian style admits of an inexhaustible
series of harmonious combinations.

I know that what I have said may not be well received by
the ordinary modern architect. I shall be told that I am flippant
and ignorant, as I have often been told before. Of this kind
of criticism I can accept any amount with equanimity. As to
my ignorance, it is my own concern. As to flippancy, I can
honestly assure any one who does me the honour to read these
pages that, so far from that, I am moved sometimes almost to
tears when I think of what is being done in all directions under
the name of architecture. A large building in a provincial town
is greatly admired there, chiefly on account of its size and cost.
I did not offer my opinion on it, but I suppose it would be
called flippancy that made me reply, when asked directly if I
admired the architecture, that I saw no architecture—only a
great deal of mixed building. The new town halls and muni-
cipal buildings in half-a-dozen English cities are in what, for
want of a better term, must be called the South Kensington
style. Architectural exhibitions are full of them and their

unmeaning details. All betray the same absolute ignorance of architectural principles, coupled with the same feebleness when attempting to obtain picturesqueness. What, therefore, with new buildings, and what with the restoration of old ones, the present prospects of sound architecture are truly deplorable.

II

THE DECAY OF GOTHIC

II

THE DECAY OF GOTHIC

"An arch never sleeps"—Progress of the Pointed Arch—Objects of Gothic builders—Flat arches—Rules—Modern imitations—The Albert Memorial—The last Gothic—The so-called "Debased Style"—Late Gothic at Oxford—The Staircase at Christ Church—The Tom Tower—Late Gothic at Cambridge—Bath Abbey—Hospital at Corsham—Charles Church, Plymouth—Hampton Court—"Peter Torrysany"—St. James's Palace—Middle Temple Hall—Inigo Jones—Wren.

THERE is an oft-quoted Arab proverb to the effect that "an arch never sleeps." Its weight is, we know, constantly acting on its supports. If they are not strong enough to resist thrust, as it is called, they are pushed asunder, and the whole edifice eventually falls to pieces. The Saracenic saying is true, in a wholly different sense, of the progress and development of English architecture. The arch of the mediæval Gothic architect never stood still. It was always changing, sometimes for the better, more often for the worse. The round Norman arch developed into the thirteenth-century lancet. The Early English style gave way to the Decorated, and both to the Perpendicular. Finally, the Perpendicular yielded to the Romanesque, and there is a return to the round arch of the Norman period, modified by the many accidents which marked the course of its long history. Between each pair of Gothic styles there was a transitional style. One kind of arch superseded another very gradually. Their inventors were constantly seeking for something, and often seem to have thought they had already attained. But they were never long satisfied. If we inquire what it was they so diligently sought, we may be led to some curious and perhaps unexpected results. The inquiry is greatly complicated by the amount of sentimental gush which has concerned itself with the origin and progress of Gothic architectural art. The answer, so far as I think I have reached it, may not commend itself to others, for it abolishes sentiment in dealing with the subject; and, if it is correct, it also gives us a reason for the failure of the modern Gothic movement or revival to fulfil the requirements of those who employ architects: to obtain, in short, the confidence of the public. The objects of mediæval builders in England were not, it seems to me, to attain any measure of mere beauty. In Early English, for example, they did their best, but they built in a particular way because they could not help it. They knew of nothing better, or we may be sure they would have tried it. They saw before them exquisite Norman work. The Galilee at Durham, the noble nave of Ely, the staircase at Canterbury, were admirable to the mind of every lover of art. But they noticed a serious defect in it all. The Norman could vault a narrow aisle, but, in the examples I have mentioned, and in many more—at St. Albans, Hereford, and Southwell—the roof was flat and of wood. There were other points which may be neglected here. The great thing was to vault a wide space. This difficulty had been surmounted partially in the little chapel of the Tower of London, where what may be called brute force was used to form a vault, the thrust of which was slight and was amply compensated by the narrow side aisles and by the enormous thickness of the outer walls. Then the pointed arch suggested itself. I do not say it was suggested, but it grew from the attempts of Norman builders at Canterbury and in a few other places, who rejected the barrel vault of the Tower, made intersecting vaults, and found that of themselves they took, in a way more or less pronounced, the pointed shape. This is the transition from Norman to Early English. It was soon found that a considerable lightness might be imparted to a vaulted building without in any way impairing its stability. The next difficulty was with the windows. Glass was rare and dear. There was no use in making openings only to close them again with stonework or shutters. So the lancet window commended itself to the designer of Salisbury, and a somewhat different shape, in which the open space was diminished to the utmost by so-called "plate tracery," found favour at Westminster and in France. It will thus be seen that, so far as we can judge, the pointed arch was first used not for its beauty but as a constructional expedient. By its use arches of different spans could be made of the same height. The first pointed style naturally gave way, when men could build better and had better materials and cheaper glass, to the Decorated style, and that again was superseded in England by the most distinctively English style of all, the Perpendicular, and in France by the Flamboyant. Both, as we shall see, lent themselves easily to the return of the Romanesque, the renascence which eventually brought in the Palladian. There is thus a succession, without a missing link, between the almost savage power of the Tower chapel through Salisbury and Westminster to the chapel of Henry VII. and to St. Paul's Cathedral.

But I have not done with the mediæval builder. I have

endeavoured to show that the pointed arch was, in a sense, forced on him by circumstances. With it only could he fling aloft those soaring vaults in which he so delighted. There are Norman arches in some places, as for instance at Norwich, as tall as the arches of Westminster or Salisbury, but the pointed vaulting which crowns them only dates from the fifteenth century. The builder of pointed arches and pointed vaulting did not use the style only because he liked it best, but because he had no choice. Mr. Stevenson has well said (*House Architecture*, i. 131) that "by breaking the round arch into two parts, attached by a point at the top, the arch could be widened or narrowed like a pair of compasses." When the builder had a choice, he speedily availed himself of it. That he did not, at any rate, always prefer it, is, I think, proved by a consultation with one of the sister arts. I have gone through an immense number of illuminated manuscripts of the thirteenth, fourteenth, and fifteenth centuries without finding a single pointed arch represented, except here and there by an accidental attempt to obtain a perspective effect, or where the narrowness of the space to be occupied has forced the artist to point his arcading. It is quite evident that the Early English artist had no special bias in favour of Early English architecture, as we call it. He thought he was building in the Roman style. But further, the tendency of every change, of every transition, was not to narrow the vaulting or to make the arch more and more pointed. Quite the reverse: every successive architect endeavoured to make his arch wider and wider, and to depress the point more and more. At last he reached the cloisters of Canterbury Cathedral. The point has nearly disappeared. One step more: at Gloucester it is gone. The place formerly occupied by the pointed head of the vaulting is flat. In the chapel of Henry VII. there is no example of a really pointed arch: the architect has attained the goal at which every successive architect from Poore at Salisbury to Bray at Windsor had aimed. He could build a roof as flat as the timber roof of the Normans. It has been well remarked, " The pointed arch had been gradually flattened till it became a straight lintel." There is a typical example at South Wraxall. We may ask why this had not been done long before. The answer, I suspect, lies in several circumstances with which we need not trouble ourselves now, such as improvements in masonry, in masons' tools, in quarrying, in means of transport for large blocks, in machinery, in canals and roads. But the moment the new Romanesque was presented to his mind the architect turned to it with avidity. It spread so rapidly that in fifty years, or less, it had taken nearly all England back to the place from which it had started three hundred years before. The exceptions, as we shall see, were chiefly in the Universities, and a few examples of great interest also occur in places where—as in Wiltshire and Somerset—the best building stone was to be found. But unquestionably the return of Romanesque was the proximate cause of the decay of Gothic. The arch which had never slept was finally put to rest. The Gothic tradition slumbered, and slumbers still, notwithstanding the efforts of the modern so-called Gothic school to awaken it.

An art which has ceased to be progressive has ceased to live. Had a mediaeval architect been told of the "rules" of Gothic architecture, he would have directed all his efforts to breaking them. Whatever it might cost, he was determined to obtain the greatest effect he could with the materials at his disposal. He went on from strength to strength. He never

turned back. For him there was only one style, namely, the
best. There was only one rule, namely, to make his building
stable. There was only one limit, namely, material. He had
to cut his coat according to his cloth. That any one with the
wealth of material at our disposal should turn his back on
progress and improvement, and deliberately set himself to build
as they built, say, in the thirteenth century, would have been
to him a thing incredible. I saw, the other day, a street front,
the upper part of which, three storeys high, some twenty-five
feet from the ground, was highly ornamental, in banded brick
and stone. The whole structure, from the aforesaid twenty-
five feet upwards, rested on an iron girder, supported only at
the ends. The empty space below was waiting for a shop
window of gigantic size to be prepared for it. Let us suppose,
for an instant, that Robert Poore, and William the Englishman,
and Sir Reginald Bray could have had cast-iron girders, to
what a noble use they would have put them. Their edifices
would have rivalled the pyramids. As it is, the modern architect
has been able to make nothing of them, except to build a few
such monstrous structures as the Prince Consort Memorial in
Hyde Park. Any one with eyes can see that not only does
the memorial not look as strong as it is, but it looks as if it
could not stand an hour. There is an enormous and weighty
canopy, in gables, surmounted by metal roofing, carrying a lofty
bronze spire ending in an immense gilt cross. This canopy is
supported on four attenuated columns, made of granite in two
colours, as light and slender as if they only held up an awning.
The whole thing would fall to pieces instantly, but for the iron
girders concealed within the masonry of the canopy. The
result is most unfortunate. The cross may stand for many

SOUTH WRAXALL.

years; as long, in fact, as disintegration from damp or any other cause does not affect the iron. Thus, then, the modern Gothic architect conceals his construction, builds a gigantic sham, and forgetting that, as I have said, the true Gothic arch never slept, really or metaphorically, he expects his wonderful building to be described as in the Gothic style, and to be admired as an effort of architectural mesmerism. It may be engineering, but I deny that it is architecture. A true architect would have said that, as such a building would and must look unsafe, at least a semblance of buttresses, even though they might not be at all needful, should be added, if only to prevent the sightseer from being afraid to approach.

A few examples of this kind show how futile it was to try to revive a style that depended entirely on circumstances which have now ceased to influence architecture. When a little engineering can make such a structure as the Albert Memorial stand upright, and when a vault can be built of any required width or height, the necessities and principles of the thirteenth century are no longer active forces. It is as utterly hopeless to try to revive them as to try to revive the thirteenth century itself. But if our architects, on discovering what those principles were, had applied them to the improvement of their own art, who can tell what they might have done? Instead, they have chosen to imitate the old work, to forget that "the arch never sleeps," and as a consequence, which most of us are inclined to regret deeply, they have discredited Gothic in the minds of the general public, while they have themselves made no progress in originality or skill.

The old succession of Gothic architects did not die out until the modern so-called revival had actually commenced. Perhaps the very last work not of the nature of a restoration which was designed according to what was left alive of the true Gothic tradition was the north transept of Westminster Abbey. It was very recently destroyed, under the name of "restoration," a name of such terrible meaning in the history of architecture. It was dated 1722, and was built under the influence of Sir Christopher Wren. The architect employed is generally said to have been a pupil named Dickenson, but Wren expressly mentions the design as his own. By 1722, the year before Wren's death, he saw it finished. The western towers, finished in 1735, show that the flicker of twelve years before had finally died down and gone out—in London, at least; for about the same time Hawksmoor was doing excellent Gothic work in Oxford at All Souls'. It is not difficult to show that the work at Westminster was carried out on the old tradition. If we take Cardinal Wolsey, in Hampton Court, and in the Hall and some other buildings which he designed at Christ Church, Oxford, as the last professor of Gothic before the irruption of the full-blown Italian or Palladian style, we can follow the succession to Inigo Jones, Wren, and Hawksmoor without a break. At Cambridge the old style also flickered long, but its influence is less clearly marked. At Oxford, the dissolution of the monastic orders, and the promulgation of the reformed religion, found many great Gothic designs actually in progress; and gave an impetus to others. Wolsey's college was not complete at his death in 1530; but one architect after another carried on the work, some of them, no doubt, great ecclesiastics, to whom the old traditions came through Wolsey from William of Wykeham, as part of the liberal education of the day. Wadham College continued and

SALCON, SOUTH WRAXALL.

improved the buildings bequeathed to it by the Austin Friars; but the typical example is Brasenose, where the pointed and round-headed arches appear side by side; and most picturesque effects are produced, in a kind of transitional style which certainly does not deserve the epithet of "Debased" bestowed on it by Mr. Bloxam and other respectable authorities. The interior of the chapel, though much "restored," is particularly pleasing and picturesque. Much of this college dates from the reign of Henry VIII. and his immediate successors, before the end of the sixteenth century. St. Alban's Hall, adjoining Merton College, was built in 1600, and is a good example of this delightful style. Oriel is a little later. The Bodleian, in its several parts, is of a very composite character, much of it having the old Gothic feeling, with a few pointed, but more round arches, here and there breaking out into Classical, but preserving a unity in its delicate beauty not to be surpassed. The architect chiefly employed seems to have been Thomas Holt of York, who designed much else at Oxford, where he died in 1624. University College is mainly in the same style, but a little earlier; for the tower over the entrance is known to have been built by the master Hamsterley, while Henry VIII. was still on the throne. Some further building was in progress in the reign of Elizabeth, much of it with pointed arches.

A most interesting paper on the late Gothic of Oxford was read in 1850 before the Royal Archæological Institute, by Mr. Jewitt, who added a list, arranged chronologically, of those buildings whose date he was able to verify. He is certainly not too enthusiastic when he characterises some of them as "highly picturesque," and adds that the occasionally incongruous details "produced great richness of effect." He specially names and illustrates the Bodleian Library, which I have spoken of above. The mixture of styles is very apparent in the gateway of the schools; but Mr. Jewitt judiciously observes that when every other trace was lost, the windows still retained their Gothic form, or their Gothic tracery.

The inner quadrangle of Merton College is said to have been built by Bentley, one of the builders of the schools. The chapel of Exeter College, built in 1624, contained some beautiful windows; but it was unfortunately pulled down in favour of a design by Sir Gilbert Scott in 1868.

Among the best known of these late Gothic designs at Oxford are the staircase and entrance gateway at Christ Church and the garden front of St. John's. The staircase is, in fact, one of the most admired buildings in the whole University. Peshall has reported that it was designed by a London architect named Smith. I cannot help hazarding the guess that by Smith he must have meant Jones. Be that as it may, the staircase, with its tall slender pillar and fan-work vaulting, was certainly built in 1640, for Dr. Samuel Fell, who was then dean. We know that Inigo Jones was building at Oxford in the Gothic manner in 1635; and the beautiful garden front of St. John's, with its exquisite oriel windows, is certainly his, and was designed for Archbishop Laud. We shall see, in enumerating the London buildings of this style, that, though perhaps not so authoritatively, the names of the archbishop and the architect were connected there about the same time. It seems strange that Smith of London should have been practising in a method so exactly like that of Jones of London, at the very same time, or within four years, and both at Oxford. It should be noted that Peshall calls Smith an "artificer," not an architect. In which case the design might be Fell's; or a supposition may be true which was hazarded, to the effect that Wolsey left a design which Smith, at Fell's instance, carried out. But the style of the vaulting is very different from that Wolsey used in the adjoining church, where there are groining ribs, as in the chapel at Hampton Court. There are, strictly speaking, no ribs in the staircase. We find fan-work vaulting at Hampton Court, but it is not of Wolsey's time; and even if Wolsey had left a design, it would have been carried out in the style of the time of Fell. The probability that he left working drawings which Fell's artificer could use seems to me extremely slight; and judging by Inigo Jones's work at St. John's and other fragments of Gothic at Oxford in which Wolsey could not possibly have had any hand, it is safer to come to the conclusion that Fell, or Smith or Jones for him, designed the Christ Church staircase.

Building seems to have been almost or quite at a standstill at Oxford during the Civil War and the Commonwealth. True, the chapel of Brasenose was founded in 1642, but it was not finished till after the Restoration. The Library was opened in 1663, and the chapel of University, begun in 1639, was consecrated in 1665. The fan-work in this college is somewhat like that of Christ Church, and must be of about the same time or later, thus helping us to bridge over the period which elapsed between the building of the staircase and that of the "Tom Tower" and the entrance gate.

This beautiful building, one of the glories of the University, has never received the admiration to which it is entitled. Regarded from the historical point of view, it is interesting as one of the last attempts at Oxford to carry on the old Gothic tradition which Wren had inherited and had strengthened by his practice on some of the churches of the burnt city of London. From the artistic point of view, it is equally worthy of notice. It is, in fact, almost impossible, as we view it, to avoid the perfectly futile speculation as to whether Wren might not have designed a still grander dome of St. Paul's in the style which he chose for Christ Church. Naturally, this building does not appeal to critics of the modern schools of Gothic. It is no imitation of the thirteenth or of the fifteenth century. But it is what so few of the architects of the past fifty years have succeeded in grasping—an application of the strictest rules of harmony and proportion to the Gothic form carried forward and upward by a genius in its maturity which, as we shall see, even in its beginnings " touched nothing that it did not adorn."

There is a place in all great art, whether that of Raphael, of Handel, or of Wren, to mention only three mighty names, in which genius and experience meet and coalesce. Sir Gilbert Scott or Mr. Street could not admire a building at once so purely original and so obviously oblivious of the rules and regulations which they had laid down as necessary to the Gothic style. But Wren perceived what they did not : that art, to be living, cannot stand still—that, in short, as we have so often had occasion to remark, the arch never sleeps. It is curious to read the following sentences in some very pertinent remarks by the late Mr. Burges on the sister art of heraldry. They are in the volume of *Gleanings from Westminster Abbey*, where he observes that in the thirteenth century people were hardly so particular about the details of a shield of arms as they became at the end of the fifteenth century "*when, like other arts in a state of decay, it became a science.*" It is curious that the writer never perceived that in this sentence he condemns the methods of the whole school of architects of which he himself was such a brilliant member. The "Tom Tower" was built expressly for the reception of the old bell of Oseney Abbey, called "Great Tom of Oxford," and is in every way appropriate to its purpose. It is octagonal in plan, with an ogee cupola of charming proportions, and is as much a feature in every view of Oxford as the tower of Magdalen, and the dome of the Radcliffe Library.

One other building in the Gothic style is almost as conspicuous as the "Tom Tower." It is considerably later, but shows Wren's influence strongly. This is the double tower and front in the north quadrangle of All Souls' College by Hawksmoor. As we enter the college, we are welcomed by a charming little domed porch in a style which can only be described as that of Wren at Christ Church. The two towers and the adjoining buildings may well have received the approval of the master, as they were in full progress several years before Wren's death. Ingram, writing about sixty years ago, is quite enthusiastic about them. "Nothing," he says, "can exceed the astonishing effect produced by the assemblage of so many striking objects as are here blended in one magnificent, though not harmonious whole. Many of its component parts will not bear criticism; yet who can stay to criticise them? The graduated stages of Hawksmoor's diminishing turrets, together with other characteristics, exhibit a fantastic air of continental Gothic; but they seem to disdain all comparison, and to stand in unrivalled stateliness, challenging our admiration." Undoubtedly, these towers and the adjoining buildings form a worthy, and indeed splendid conclusion to the long list of fine Gothic buildings at Oxford.

It had not lingered so long at the sister university. The tower of "Great St. Mary's" at Cambridge was designed by

John Warren, in or before 1608. It has little merit, though it is 131 feet in height. The chapel of Peterhouse was begun in 1628 by Matthew Wren, afterwards Bishop of Ely, the uncle of Sir Christopher. The curious gable ends, with a kind of pediment, contain very simple perpendicular windows, the general effect being unquestionably picturesque, and very superior to the mock Gothic of Gisborne Court, built in 1825. The grand chapel of King's College was not complete in 1524, when it would seem that Italian workmen were introduced; and the wood carving shows strong traces of their handiwork, affording the same picturesque effects of contrast as we see in parts of the chapel of Henry VII. at Westminster. Some further work in panelling, much more Gothic in design, was put up as late as 1595.

Some of the most interesting buildings in the slowly dying style are naturally to be found where good, easily carved oolite abounded. Great Chalfield Manor-House is in genuine Perpendicular Gothic, having been built in or about 1490. The manor-house of South Wraxall, already mentioned, a couple of miles off, shows in its purely Perpendicular gateway an example of a perfectly flat lintel. The Gothic tradition was further kept alive in this neighbourhood by the gradual completion of Bath Abbey, as it is called. It was in hand in 1499, but was not finished till 1616; the style being preserved throughout without deviation. It is too uniform, too featureless, to be very interesting. So completely, however, did the architect get rid of any feeling of the necessity of pointing the openings that the great east window is actually square, or, as it is sometimes termed, cottage-headed, though it is 50 feet high and 20 feet wide.

ALMSHOUSE, CORSHAM, WILTS.

One of the latest buildings in which the Gothic tradition ruled the design, is in the same oolitic region. This is the Hungerford, properly the Halliday Hospital, a little almshouse with a chapel, at Corsham, a few miles from Bath. It contains no pointed arches, but is purely Gothic. The windows of the chapel have mullions, and an attempt at tracery, and over each is a dormer, which has a singular effect. A similar and very interesting effect was produced by the building of a side aisle and gallery in the parish church, but was scrupulously removed at the "restoration" a few years ago. The porch of the hospital has an attempt at Ionic pillars on either side of the doorway. Above, and also on the north front, are shields of arms, and an inscription from which we are surprised to learn that the date of the building is 1668. The almshouse was founded by a widow who is described as "Lady Margaret Hungerford," and who evidently intended it more as a memorial of her father and her grandfather than of herself or her husband, Sir Edward Hungerford of Corsham Court. She was the daughter and coheir of William Halliday, or Hollyday, an alderman and sheriff of London in 1617, by Susan, his wife, daughter of Sir Henry Rowe, or Row, Lord Mayor in 1607. The arms of Halliday are represented above, with the crest and a little shield of Hungerford higher up still. The Halliday arms—three helmets, and the motto, "Quarta Salutis"—are an excellent example of Jacobean heraldry.

The whole of this region abounds in old houses and churches containing monuments worthy of the architect's study. It is common to fix a date for the discovery and introduction of oolite as a building stone; but it was used here from time immemorial; and here, as we have just seen, the Gothic tradition continued to linger even after the outward form had been wholly changed.

At Fulmer, in Buckinghamshire, there is a Gothic church, built in 1610, of brick, which shows that even where there was no stone the old custom still obtained.

A very interesting church is at Plymouth. It is generally said to have been consecrated in honour of "the Blessed King Charles the Martyr," but this is a mistake. The parish in which it stands was separated from St. Andrew's by Act of Parliament in 1640, and designated, after the name of the reigning King, "the parish of Charles." The church is therefore not "the church of the Blessed King Charles," but "the church of the parish of Charles." It bore this name all through the Civil War and the Commonwealth down to the Restoration. It was completed in 1657, according to an inscription on the tower, so that we have here an example of a building in the Gothic style commenced on the eve of the Civil War and carried to completion under the rule of the Protector. The authorities of Plymouth are not worthy of the interesting church in their care. A determined attempt has been made during many years at intervals to Gothicise its features, and, unfortunately, with but too much success. It is, therefore, difficult to obtain a view in which its true merits may be fully discerned. The proportions are more carefully calculated than is usual in ancient Gothic work, and still less in its modern imitation, the relation of height to width in the nave being especially satisfactory. Most of the windows have had pointed openings forced upon them, but this is not so apparent within as without. A row of arches, only one of them pointed, divides the nave and side aisles, and the roof, in a barrel-shaped ceiling, with ribs of dark oak, accords well

CHARLES CHURCH, PLYMOUTH.

with the delicately carved fittings. The mouldings of the arches are much better, in undercutting, than is usual in the modern mock Gothic style; and the whole design, so far as we can still judge of it, was evidently well thought out by an architect who deserves better of posterity than to have his name forgotten, and his work handed over to some modern Vandal, who thought he could improve it. At Tavistock, not far off, is another church, with an aisle in the same style dated in 1670. There are three aisles in late Perpendicular, and the fourth, with interesting carved granite capitals, round arches, and two south doors, is strictly Gothic, in spite of the shape of the arches.

Nearer London, the later Gothic was more common, but much of it has disappeared. Some has been destroyed bodily. Some has been "restored" out of knowledge. Beginning with the period of the suppression of the monasteries and the beginnings of Italian art in England, we may visit Hampton Court, where one of the finest and most typical of Gothic buildings is the Great Hall. Here Italian details are numerous, but the whole design does not differ materially from that of Eltham and other halls of earlier date. Mr. Law, in his *History of Hampton Court* (i. 155), shows that the architect in all probability was Henry Williams, a priest, who was "surveyor of the works." Mr. Gotch, in a paper read before the Architectural Association in March 1892, has proved that "surveyor" meant the same as architect, and shows that this is the Shakespearean use of the word (*Henry IV.*, pt. 2). It subsisted at least until the time of Sir Christopher Wren. The strong similarity of the hall at Hampton Court and that at Christ Church suggests that they are both by the same architect. The hall at Christ Church is dated by the best authorities, including Mr. Papworth (*Renaissance and Italian*, p. 6), in 1529; while that of Hampton Court is known for certain to have been commenced in 1530. To build it, an older hall by Cardinal Wolsey was pulled down. The probability seems to lie in identifying Williams as an architectural pupil of Wolsey, and conversant with the principles on which his master would have planned such a building. The details, which are very Italian at Hampton Court, and less markedly so at Oxford, would be left, as Mr. Gotch has pointed out, very much to the taste and fancy of the artists employed. This would be in strict accordance with the Gothic tradition as I have endeavoured to state it above; everything being subordinated to the supposed necessity of obtaining that which was thought to be best, irrespective of style; for critics and historians of art did not then exist, and the possibility of two incompatible styles causing a difference of opinion had not dawned on anybody's mind. We see this state of ignorance and innocence, this delicious unconsciousness of the birth of questions which were destined for some three centuries to agitate the minds of the votaries of taste, in the thoroughly Italian work of Torregiano, surrounded by the Gothic grill of bronze by an English artist, which form together the central feature of the chapel of Henry VII. "Peter Torrysany," as his insular employers called him, also made the beautiful but wholly incongruous monument of John Young, Dean of York, who was Master of the Rolls, and was buried in the Rolls Chapel. The chapel itself, which dates in part from about 1516, presents no features of interest; but John Young's tomb is one of two or three worth seeing. The chapel of St. Peter "ad Vincula" in the Tower is another example of about the same date, it having been rebuilt after a fire in 1512. Here any characteristic architectural features have been scrupulously removed by successive "restorations." The Chapel Royal, Savoy, was also built in the early years of Henry VIII., and is distinctly Gothic in style; but a fire and a "restoration" have left very little of the old work visible. The old Gothic gateway of Lincoln's Inn is still standing, though oft condemned. It was built in 1518, and has a slightly pointed archway. This, and St. James's Palace and St. John's Gate at Clerkenwell, built in 1504, are the only domestic Gothic buildings left in London, though some parts of the Guildhall may be older.

St. James's was designed as a kind of hunting-lodge, an outlying appanage of Whitehall, which itself was made by Act of Parliament part of the palace of Westminster. As palaces, both Westminster and Whitehall have ceased to exist; but St. James's preserves many Gothic features, chiefly external. In the Presence Chamber, a fireplace bears the initials of King Henry and Queen Anne, so that it must have been built between January 1533, when they were married, and May 1536, when Anne was beheaded; if indeed the initials do not stand for those of Anne of Cleves, which would make the date 1540. This date, with the same initials, is painted on the curious panelled ceiling of the chapel. There is a legend to the effect that both the chapel and the tower which fronts St. James's Street were designed by Thomas Cromwell, the Vicar-General, who was made Earl of Essex in April and beheaded in July of that same year, 1540. This is also the date of the Horse Shoe Cloisters at Windsor. Hans Holbein designed much for the king at Whitehall, but whether he had a hand in any part of St. James's it is not easy to say, and he died in 1543. From this time to the end of the century, very few buildings in pure Gothic claim our notice. The style was evidently out of fashion. The age of John of Padua, of Shute and Thorpe, had come; and in London we can only now point to the hall of the Middle Temple, where, chiefly in the plan, the Gothic tradition survives. But the screen, and nearly all the details, are Italian in character, though, as at Oxford, Perpendicular tracery was still used in the windows; the older fashion lingering longer among the glaziers and lead-workers than among the wood-carvers. Glass was still comparatively scarce and dear, and was made in small pieces, so that leaded lattice-work was a necessity. The Middle Temple Hall was finished in 1572, but after this, for many years there is no record of a Gothic building being erected in London. A few examples occur, as we have seen, in the country; and the old tradition, though dying, was by no means dead. It lived in the oolite regions of Dorset and Wilts, and was active at Oxford, and in a minor degree at Cambridge. In many places there are examples where the old style enters into competition with the new, as in the chapel of Bishop West, 1534, at Ely. The chapel of Bishop Longland at Lincoln may be dated at 1547, but was never finished. Signs of Renascence work are very apparent in the rich ornamentation. The fall of the religious houses had, no doubt, a discrediting effect on the architectural style in which they were built; but a temporary reaction occurred, and it can hardly be doubted that Inigo Jones actually preferred the Gothic at one period of his career. He had travelled on the Continent, and we find him in Denmark in 1589. When the Princess Anne came from that country to wed King James, she is said to have brought him to Scotland with her; and critics see his hand in Heriot's Hospital at Edinburgh. But before Heriot's can have been even begun, Jones had left Scotland; and we find him back in Denmark, whence, in 1603, he again seeks the English court, and is appointed architect to the Queen, for whom he designed the scenery of masques, among other things. He was in Italy before 1604, and at Oxford in 1605; but the dates of his two chief Gothic buildings in London must be placed considerably later. The chapel of Lincoln's Inn was much as he left it, until 1791; when it was "restored" by Wyatt. The cloister underneath was still intact in 1882, when it was handed over to the mercies of a wealthy amateur, who employed a Mr. Salter to remove from it as far as possible the traces of Inigo Jones's hand. A description of the building by Mr. Weale is in his *New Survey of London* (p. 175), published in 1853: he says of it that "the side elevation of the exterior plainly partakes of the boldness, stateliness, and harmony of his other designs; and though," he continues, "the petty exactness of later imitators may yet find it convenient to make faults of every variation from precedent in the details, this fragment has some rare qualities. We know of no mediaeval work, even, in which apertures of so low and broad a proportion produce, as here, no ungraceful or mean effect." Of Inigo Jones's other London Gothic work, Mr. Weale has not so high an opinion. This is the church of St. Katharine Cree in Leadenhall Street. I unhesitatingly ascribe the design to this architect. He had already designed for Bishop Laud at Oxford; and though, no doubt, the Bishop would dictate the main features of the plan, the rest of the design would be left to the architect. Laud may have been attracted to him by his religious views as well as by his genius; but it is scarcely doubtful that the church is mainly his. St. Alban's, Wood Street, was rebuilt by Jones, but perished in the Great Fire, when it was rebuilt a second time by Wren. It has been sometimes suggested, and with plausibility, though without proof, that Wren's design was identical with that of his predecessor. Like the chapel of Lincoln's Inn, it has suffered much at the hands of the modern Vandal, and is now hardly worth a visit.

Of Wren's Gothic work in London, it will be time enough to speak further on. Suffice it here to say, that it has the same sincerity of purpose, the same originality, the same straightforwardness as his Palladian work. That he or any other architect, conscious of the slightest creative faculty, should be content to sit down, turning his back on the state of the arts and sciences of his day, and deliberately endeavour to design in a style which had worked itself out in the thirteenth century, would have been a thing altogether incredible. It is true that Wren imitated the thirteenth-century architect in one point. He tried to do his best with the materials at his disposal. It has been reserved for the taste of the present day to destroy his last, and in some respects greatest work in the style; but his towers of St. Michael, St. Dunstan, and St. Mary Aldermary, in the City, have of late been recognised as masterpieces; and it must be the endeavour of all lovers of what is good and progressive, not imitative and retrograde in architecture, to insure their preservation for the instruction and pleasure of generations who will look upon the short reign of mock Gothic as a period of decay, falsehood, and destruction.

III
ELIZABETHAN ARCHITECTURE

III

ELIZABETHAN ARCHITECTURE

A Time of Change—A new Style—Elizabethan Houses—Examples—The Irregular Type—The Regular Type—Haddon Hall and Longleat—The Duke's House, Bradford—Cheshire Houses—John of Padua—The Masons—Lord Burghley.

WHILE the old Gothic style was flickering out in England, architects found themselves very much as they find themselves now. But, unlike the present, that was a great creative age. Originality was to be seen everywhere. The poets of the day were Shakespeare, Spenser, and a legion of competitors who have made so famous those "spacious times." It was the same in art. True, Holbein had died in 1543, but he had founded a school, and its efforts were aided and directed by the foreign artists imported. It was the same in music. We have never heard sweeter strains than those first sounded by Merbecke, and Tallis, and Lawes. It was impossible that architecture should lag far behind. By one of those curious coincidences which occur in the history of art as well as in that of any other movement, while we were most busily engaged in cutting ourselves adrift from the religion of Italy, we were most diligent in studying her architecture. I have endeavoured to show that at first there was no necessary antagonism between our old Gothic and the new Palladian. In English buildings every attempt was made to reconcile them, to adopt what was best in each, to construct what would be at once picturesque, correct, and suitable for a climate very different from that of Italy. The result was extremely satisfactory. Haddon Hall, the palaces of the Cecil family, the rooms now used for the Royal Library at Windsor, and many more of the "stately homes" of which Englishmen are so proud, were designed and built at this epoch. Mr. Gotch lately read a paper before the Architectural Association, in which he endeavoured to answer the question of how they built in the time of Shakespeare. He approached the subject from a purely architectural point of view, and came to definite conclusions as to surveying, planning, designing the elevations, "rating the cost," superintending details, and all the other items which went to make up a beautiful house. I shall have further occasion to refer to this brilliant essay, and can only hope that the architects who heard Mr. Gotch may lay some of his tacitly implied counsels to heart.

The names of the greatest architects of the day have come down to us in sufficient number to enable a critic to form an opinion as to the whole school and its style. Nothing valuable in the old Gothic tradition was violently thrown aside. A more than semi-Gothic hall existed still in every great house. Many other features which had descended directly from the thirteenth century were retained. In some houses, such as Haddon, the old irregularity of outline was not looked upon as any defect. In others, as at Longleat, a façade almost Italian imparts stateliness at the cost of the picturesque. Architects were still afraid to use the classical forms exclusively, and the symmetry of Hatfield or Cobham was not produced in a day. There was, however, no affectation about their work. Window glass, as I have had occasion to point out, had a strong influence on the shape and size of windows. At the present day, we can have a single pane large enough to fill a whole window, of whatever size. Then, the beautiful lattice work which was part of the old Gothic style was still a necessity. The great hall, similarly, was retained as convenient for a great household. At Knole, and no doubt in all other houses of its class, there was at mid-day a dinner to which every one, from my lord to "my lord's blackamoor," sat down. A long gallery was another necessity. It was to be sunny and light, as at Haddon, giving easy access to the garden in summer, serving as a place of exercise in winter, and well warmed with more than one great fireplace, as at Cobham and Loseley. There was found to be less draught from the windows and doors of a large chamber than from those of a small one; and so all the reception rooms and many of the others were of vast size, as at South Wraxall, where a comparatively modest house has at least three or four chambers as large as the whole of a modern villa. The great defect of all was the poor bedroom accommodation. It was no unusual thing for the whole of what we should describe as private rooms to be accessible not from a passage, but simply through other rooms, so that to reach the farthest of a suite the rest had to be traversed. This was in accordance with the manners of the day. There was no privacy. Every one lived more or less in public. Men, except at court, kept their hats on. Women wore hoods and head-dresses. In bedrooms without fireplaces warmth was insured at night, no doubt, by retaining a portion of the clothes worn in the day; and it was no unusual thing for two, three, or even more persons of the same sex to occupy one bed. Of this coarseness of manners we have abundant evidence, into which it is quite unnecessary to go here. The life, as portrayed by Shakespeare, by Spenser, by Herrick, even down to the comparatively civilised times after the Restoration of the Stuarts, was of a character very different from anything of which we have experience at the present day. Our modern houses, however inconvenient they may often be, contain a thousand items of comfort then undreamt of and unknown. Magnificent and beautiful as some of these old palaces are, it is only by judicious alterations that they remain habitable. When Wyatville went to Windsor Castle, he found a series of towers connected by a curtain wall, much as we see the Bell, the Beauchamp, and the Devereux Towers connected in the Tower of London. All were "passage rooms." Though some were wide and high, all were inconvenient; and the royal family lived, as described by Madame D'Arblay, in a house in the garden, and only used the older chambers on state occasions. Wyatville, by building the long corridor within the line of towers, made Windsor Castle a modern abode. Something of the same kind has had to be done in many other old houses of slighter pretensions; and the adaptation of Longleat, of Knole, of Holland House, and of Cobham Hall, to name only a few, has sometimes been a very difficult matter, and tried the ingenuity of the best architects. Two or three features are common to nearly all Elizabethan houses. There is always, as I have said, a hall, a gallery, and a withdrawing room or solar. Most of them have also a chapel. Even in comparatively small houses like Ightham Moat, this, or at least an oratory, is not omitted. At Knole, the chapel is very large, and has a great gallery, approached by an antechapel at a higher level. At Hardwick, the chapel is very small; it is larger at Hatfield and Burghley, but appears to be altogether missing at Cobham. It may have been an early feature of the house. There is a good chapel at Westwood, another at Longleat, and one was at Audley End.

The design of some of these great houses is highly irregular. This is the case at Knole, Penshurst, and especially at Haddon, where we must remember that an ancient house was added to in 1540 and the succeeding years down to 1589. But Westwood, Hardwick, Longford, Cobham, Hatfield, Burghley, Audley End, Eastbury, and many more are perfectly regular, the ground-plan forming often a letter E or a letter H. Longleat seems to have been one of the first important houses built in this fashion. It was in progress from 1567 to 1578. The beautiful "Duke's House" at Bradford-on-Avon belongs to the same style and period, and is usually ascribed to the same architect, who may have been John of Padua, or even more likely, John Thorpe.

It may be worth while to examine more closely some examples of each kind; the irregular naturally coming first, as they are usually founded on older buildings. A favourite specimen is Haddon Hall, which rises on a steep slope above the Bakewell meadows, and is first seen among old trees and across green grass. The contour of the ground materially influences the design, both in elevation and in plan. It has, besides, more of a castellated appearance than other buildings of the same period. The entrance is under a tower at the north-western corner and admits the visitor to the first, or lower court. Everywhere there are flights of steps,—steps up and steps down,—and the picturesque appearance of the building is greatly enhanced at the expense of its convenience. This lower court was never intended for the admission of a carriage; and had the house remained a family residence, a new main entrance would have had to be made, as in other places. Probably this part is older than the Elizabethan period. A fortified manor-house was on the site long before. The license to crenellate is dated, according to Mr. Papworth, in 1199. Sir George Vernon, to whom the greater part of the existing building is attributed, came into the estate as early as 1515. He was more or less actively at work on the house down to the time of his death in 1567. The work was completed by his successor in the ownership, Sir John Manners, who is said to have eloped with Sir George's elder daughter, Dorothy. Be this as it may, the existing "stage scenery" of the event, as shown to visitors, where the fair Dorothy is made to leave the gallery, or ball-room, by the anteroom, and so pass out to the terrace, cannot be accurately delineated, for the simple reason that this part of the house was built by herself and her husband after they had long been married and had inherited the estate.

This great gallery is now one of the principal features of Haddon Hall. It runs west to east along the south side of the upper or second court, and contains two bows and a rectangular bay window 15 feet by 12, all deeply recessed, so as, in fact, almost to make three separate apartments. The wainscoting shows the gradual approach of the Palladian style, being wholly different from that in the older parts of the house, where Gothic still asserts itself. Something very like a composite capital crowns each pilaster. The ceiling is of plaster covered with a beautiful pattern of scroll work with heraldic shields, formerly, no doubt, coloured and gilt.

The garden, with its terrace, is one of the chief beauties of

"GALLERY, HADDON HALL."

Haddon, being a combination of the formal and the natural which commends itself to both schools of gardeners. The arched balustrading is very characteristic of the Elizabethan style, while recollections of the Gothic practice had not yet been wholly obliterated by the Italian. Mr. Blomfield (*The Formal Garden*, p. 114) says of the terraces at Haddon that they are laid out in four main levels: "At the top is a raised walk 70 paces long by 15 wide, planted with a double row of lime trees. About 10 feet below this is the yew-tree terrace, divided into three plots, about 15 yards square, surrounded by stone curbs, with yew trees in each angle. These were once clipped, but are now grown into great trees overshadowing the entire terrace. Dorothy Vernon's stairs descend on to this yew-tree terrace. A flight of twenty-six steps led from this terrace to a lower garden, about 40 yards square, divided into two grass plots. A walk from this garden skirted round two sides of a second garden laid out in three levels, and reached the postern door in the outer garden wall by seventy-one steps laid out in seven entire flights." With regard to the balustrading of the terrace mentioned above, Mr. Blomfield has some interesting notes: "The terrace at Haddon has six small stone arches to each bay. The height is 3 feet; width from centre to centre of piers 11 feet 6 inches; the steps measure 12 inches by 5."

Much of Knole partakes of the irregular character of Haddon, the more formal parts appearing, like the oriel window in the first or Green Court, to be purely Gothic. Here the irregularity was largely caused by the local absence of good building material, and some of the most picturesque parts are of wood, as for example the south or garden front. It is strictly contemporary with Haddon, and both with Penshurst, but the Elizabethan buildings at Knole are attributed by Mr. Papworth unhesitatingly to John Thorpe, on good grounds, I do not doubt. Penshurst and Haddon both look older in places than any part of the second or Stone Court of Knole. The hall at Penshurst is purely Gothic, but the entrance tower dates from 1585, probably the date of the gallery at Haddon. Haddon, Penshurst, and Knole are typical examples of the irregular Elizabethan—that which grew more directly out of the Perpendicular Gothic, and contains much of it mixed up in work of a later type. The hills of Gloucestershire and Wiltshire are dotted with smaller examples. Their characteristic features are the cross mullions, the gables, the frequently pointed doorways, and the sparing use of classical pillars and pilasters. The principal rooms, even in mere cottages, have generally high carved mantelpieces, the walls are panelled, the ceilings are of timber or of ornamental plaster, and porches are numerous, with or without small chambers above. In one ordinary manor-house of two sitting or reception rooms and not more than four bedrooms, there are five porches. The chimneys are often very ornamental, but chiefly where they are of brick, those of stone being comparatively plain. The balustrades of the staircases are at first merely rows of small arches, often very ornamentally treated. By degrees the head of the arches separate, each half-arch becoming a kind of bracket for the hand-rail. Finally, but not until the seventeenth century was well begun, the new Italian balustrade, more or less modified in its translation into oak, came into fashion. The curious visitor to one of these old houses should never neglect to look for the exterior lead-work. The Gothic gurgoyle was succeeded by piping with, often, a large shield-shaped head on which initials, dates, and even, as

JAGGARD'S MANOR-HOUSE, CORSHAM, WILTS.

at Claverton, Sherborne, and some other places, armorial bearings.

Almost side by side with these delightful rambling, irregular piles, half castle, half manor-house, there grew up a series of stately palaces, such as we see at Longleat, Loseley, Bramshill, Audley End, Blickling, Hardwick, and many other places too numerous for mention. They have very similar details to those of their irregular contemporaries, and chiefly differ in the elaborate symmetry of the ground-plan. The Gothic tradition prescribed that the chief features of a house should show on the outside. You can see where is the chapel, where the gallery, where the hall, and so on. But in the regular school it is but seldom that the visitor can say for certain that a window lights a saloon or an oratory, a ball-room or a dressing closet. To the end of the sixteenth century, and much later, these symmetrical designs seldom exhibited columns or pilasters, but they were essentially, in the modern jargon, "astylar." This treatment looked well beside either the old Gothic or the later Palladian; Inigo Jones's additions to Cobham have an effect as good as that of the remains of the episcopal palace at Hatfield.

We can easily understand the charm of the new style to the minds of our ancestors. There were still standing and inhabited hundreds of fortified houses and castles, where security was sought to be purchased by discomfort, gloomy and dark, with the smallest possible space for the living rooms. Many of the less pretentious of the manor-houses were surrounded by moats, a constant source of damp and disease. The castles disappeared for the most part in the Civil War. Very few were still habitable when Charles II. was restored. We can well believe that the subjects of the Tudors, who remembered or had traditions of the hundred years of the Wars of the Roses, were delighted to take advantage of the peace and security they enjoyed; and though, so greatly to our advantage, some new houses were planned on the old pattern, but no longer gloomy and cramped, it is easy to judge how they must have welcomed the possibility of building on a regular plan with symmetrical parts, and of making sometimes the whole wall of a great chamber or a hall one continuous window.

Among the smaller houses in this second Elizabethan style, the most perfect is that known as "The Duke's House" at Bradford-on-Avon. There are many beautiful buildings in and around this little town, and the visitor will be charmed with the old bridge, a mediæval barn and farmhouse near the railway station, a magnificent church, a wonderful Saxon chapel, several gabled houses of considerable antiquity, some excellent Palladian examples, and the Duke's House. He will also wonder, in the midst of so much that is good, to see some modern buildings in the style, or no style, which has sprung from the ruins of the so-called Gothic revival. Altogether, Bradford, standing as it does in the midst of the best oolite region, a place where cloth-working brought prosperity while as yet there was taste left among us, is a museum of architecture. South Wraxall is within a few miles. The later domestic Gothic is well illustrated by Great Chalfield, built before 1490; and, of older remains still, there are the ruins of Charterhouse Hinton, in Early English, and the George Inn, at Norton St. Philip, in Perpendicular.

The Duke's House is a little to the east of the town, on a gentle slope. The front has two storeys, topped by attics under three gables. The central window projects squarely; the side windows are much wider, and each projects in a small semi-

THE DUKE'S HOUSE, BRADFORD-ON-AVON.

circular bow. Over the windows is beautiful flat balustrading, not in the least Italian, yet not Gothic. This balustrading is typically Elizabethan; and on the terrace and steps into the garden it is of the same character but of a different pattern. Between the projecting windows are others flat, so that the whole front is taken up with a series of lights, those on the ground-floor being interrupted only by the entrance. These windows are formed by stone mullions; two transoms in each opening running along the whole front. The chimneys are plain and square, set corner-wise. There are two gables at the side of the house, with four tall, plain, double-cross-mullioned windows in two storeys. The back is very like the front, but plainer. The entrance doorway from the terrace is the only place where we see any Italian features; two graceful but very plain engaged columns standing on either side. Unlike so many houses of the period, the Duke's has no courtyard, the centre being occupied by a wide newel stair: an unusual and very pleasing feature. The rooms are, of course, magnificently lighted, and are high in proportion to their size. The ceilings are beautifully decorated with plaster fret-work. The house has been restored, but judiciously, and has been always kept in good order. There is not, except in a kind of cresting over the door, and the balustrades already mentioned, an inch of ornament anywhere; yet the effect is ornamental in no slight degree. The whole front is about 50 feet high and about 60 wide. The balustrade of the terrace is described by Mr. Blomfield (p. 185) as formed of panels of stone, $3\frac{1}{2}$ inches thick, pierced with open work of alternate lozenges and ovals, with engaged balusters to the piers, and stone urns of various designs. Of the whole terrace, he tells us that "the fall of the ground is very sudden,

but that the difficulty is got over in a very skilful way" (p. 104).
"The house is raised 12.0 above the lower garden; in front of
the house is a terrace 24 feet wide, with a flight of fourteen steps
in the centre, descending to a grass platform with mitred slopes.
The path runs to right and left, and descends to the lower
garden by flights of seven steps: off this path, on either side of
the terrace walls, two steps ascend to grass terraces 27 feet wide,
and 52 paces and 29 paces respectively, which run under the
walls of the upper gardens to right and left of the house."

It is easy to imagine the surprise and pleasure with which so
marked a departure from precedent must have been hailed when
this house was first built, and the garden laid out. There is
nothing, except the modest columns at the door, to identify the
style as more Italian than Gothic. It is, in short, Elizabethan;
and from the novelty of the plan and the suitability of the whole
design to our climate, may as well have been the work of Thorpe
as of John of Padua. We cannot believe that the architect, if
he came from Italy, would have omitted some Italian details
which we fail to find. The Duke's House was built for John
Hall, a wealthy merchant of old family in the town, between
1567 and 1579, and passed eventually into the possession of
Evelyn Pierpont, Duke of Kingston, a great magnate in the
reign of George I.—hence its name and that of Kingston House,
used by Mr. Blomfield. The notorious Duchess of Kingston is
said to have been much here; and some years ago there were
still reminiscences of her eccentricities among the townsfolk.

Aubrey, in his notes on Wiltshire, attributes the design to
John of Padua. There can be little doubt that the same
architect designed both it and Longleat. In each there is the
same reliance upon proportion, rather than upon ornament, to

LONGLEAT.

insure an ornamental effect; and though the Duke's House claims our attention first, as being easier to understand, from its small dimensions, and as being practically unrestored, Longleat, it must be acknowledged, is far more important and interesting, if only for its great size. Its chief characteristics may be briefly summed up. It forms a parallelogram of 220 feet by 180. The ground-floor is 15 feet high, the next 18, and the third 12 feet. The architecture is mainly the same as that of the house at Bradford; but of course Longleat is on a much larger scale, and has not the same charm of compact beauty. There is a similar absence of mere ornament, the same abundant fenestration, the same beautiful parapet work, and, as compared with contemporary buildings, the same freshness and originality. If ever it could be said that a style was invented and did not grow from something else, it might be said of the style introduced in these two houses. At Longleat there is a certain amount of use of more distinctly Italian features, such as engaged columns between some of the windows and at the principal entrance; but there have evidently been reparations and alterations at different periods. The older part, attributed to John of Padua and to Thorpe, was completed by Smithson; and, after a fire, Sir Christopher Wren made many additions and improvements. Wyatville also had a hand in it; and it is better, on the whole, as I have said, to accept, as typical of the style, the Duke's House at Bradford.

Longleat and Bradford do not stand alone. They are only two of a goodly company. In Cheshire and some other counties where the oolite was not to be found, the style is represented by Samlesbury Hall, Poole, Moreton, Agecroft, and others—all of timber and plaster. In Northamptonshire and several of the adjoining counties, building stone is to be had, and many Elizabethan houses still exist, while a still larger number have disappeared. Some, like Kirby, are empty; some, like Burghley, have been altered; and some, like Kenilworth, have become picturesque ruins. When we come to real Palladian architecture and the use of the orders, we shall find many examples in which an older Elizabethan house has been added to or decorated; but the abundant allowance for window space was common to all, as well as the use of gables, sometimes, as in Kent, curved, of intricately designed chimneys, of flat lace-work parapeting, and the absence of Italian cornices and engaged columns. These gables are extremely obnoxious—I do not know why—to "restorers"; and at Ingestre and many other places they have been removed.

There are a few architects' names to be noted as belonging to this epoch. We cannot always distinguish between architects, surveyors, and stewards entrusted with the carrying out of designs. Tradition has handed down the name of John of Padua. Of his work we know very little, if anything, for certain. He came from Italy in or before 1544, and on the 30th June in that year he received from Henry VIII., then at Westminster (Whitehall), a grant of 2s. a day "for services to the King in architecture and music." In 1549 there is a grant from the young King Edward VI. of the same sum; being probably a continuation or confirmation. It is easy to form a theory as to John and his career. We may suppose he came, like so many others, to Henry VIII. as a musician; and having, in common with most of his compatriots, a turn for art, he may have offered his services to the King and other English employers of architects. He may, and probably did, offer to build after the manner of Italy—an undertaking he would be unable to fulfil unless he had learnt the art at home. That he had vague recollections of the beautiful buildings rising in Venice, in Florence, in his own Padua, is likely enough; but there is a complete absence of any evidence that he was acquainted with Vitruvian, or even Palladian teaching. No such design can be found among the buildings which have been attributed to him. If, for example, he was an architect educated in Italy, why did he build Longleat—assuming the fact, for an instant—without arched or angular pediments to the windows, without an Italian cornice, and with engaged pilasters of no particular proportion and in no particular style? He deserves credit for an original design; but, as I have pointed out, it is as little Classical as it is Gothic. Apart, however, from the two entries of royal grants, and from the persistent traditions above mentioned we have no precise information. Mr. Papworth mentions his name doubtfully in connection with Somerset Place in the Strand, of which nothing remains, and also with Sion House, now completely altered, with Longleat, and with Bradford. Further than this we have nothing; and in passing on to the next name on our list, we may just pause to recall the pretty old legend that John of Padua died while sitting tranquilly in the garden of Longleat and watching the work of his masons.

Next in order, many would be disposed to place John Thorpe; but he belongs strictly to the next chapter, as do the two Shutes and Robert Smithson. Henryck is mentioned by Lord Burghley, and was probably employed to design more than a bay window at Burghley House. He had been brought over by Gresham for his Royal Exchange, and came from Antwerp; but of his further history we know nothing. There is a certain interest attaching to the "masons" who actually superintended the erection of Burghley and Cobham and other great country houses. If, as Mr. Gotch believes, the surveyor or architect did little except plot out, plan, or, in Shakespearean phrase, "draw the model," the execution was entrusted to the master mason, and under him were the carpenters, the glaziers, the makers of leaden cupolas and images, the plasterers, and the carvers. The architect made no drawings of details, and all such things were handed over to "a local agent or foreman, or clerk of the works, who hired labour on behalf of the building proprietor, overlooked the men for him, made bargains with them for doing the work, and paid them from time to time." So far, the old traditions lingered. The architect probably imparted, as far as he could, an Italian air to his general outline, to his domed turrets, and to any porches or other features which would bear a design of pillars and arches. But the Flemish and Italian plasterers, with in some cases the wood-carvers, who were also often foreign importations, would alone understand the object of the architect. The rest of the work would retain an air of the Gothic style in which the artificer was brought up; and so we have the fascinating combinations which were bound to result—combinations of skilled single-minded workmen who had learned that certain things were good, certain things bad, and that, as I have had so many occasions to observe, even thus early in my book, there was only one style, namely, the best. It is rather to this straightforward disposition of the workmen, and to the influence of a competent overseer, that we must attribute the beauty of these Elizabethan houses. At Burghley we have Peter Kemp; at Cobham, Richard Williams; and at Hatfield, Robert Liminge. One other influence must be mentioned. As Mr. Gotch observes, these overseers wrote for instruction and to report progress not to any architect, but to their employer. It is to the taste and liberality of such men as Burghley and his sons, to Hunsdon, Stafford, Tresham, and other wealthy amateurs, rather than to Thorpe, or Shute, or Smithson, that we owe the best of these ornaments of our land : the worthy exponents in architecture of the new birth which influenced poetry, the drama, history, music, and all the other arts in the great days before "the setting of that bright Occidental Star, Queen Elizabeth, of most happy memory."

IV

THE BEGINNINGS OF PALLADIAN

IV

THE BEGINNINGS OF PALLADIAN

The Beginnings of Palladian—The First Examples—Tombs by Torregiano—Sir Anthony Browne's Monument—Mantelpieces—The Royal Exchange—Caius College—Recent Vandalisms—The Gates of Humility, Virtue and Knowledge, and Honour—Palladian in Fashion—John Shute—Lomazzo—Birth of Inigo Jones—Gothic and Palladian—Palladio—Vitruvius—Proportions of the principal Orders.

STRANGE to say, it is not in houses but in churches that we must seek for the first efforts of the Palladian architects; although, strictly speaking, this, the latest development of Italian taste, had not yet taken place, and Palladio was probably not yet born. In hundreds, perhaps in thousands of cases, the fell work of the "restorer" has destroyed these early monuments: but enough remain to show us how warmly the style was appreciated, and how correctly it was practised before any house or church had been built in it in England, and while in Italy the style was still in its infancy. The first examples are the three tombs by Torregiano: two of them in Westminster Abbey and the third in the Rolls Chapel. Henry VII. and his mother, the Countess of Derby, generally known as the Lady Margaret—"the Lady" being precisely equivalent to our phrase "the Princess"—both died in 1509. John Young, Dean of York, was made Master of the Rolls in the beginning of 1508, and continued in the office by Henry VIII. on his accession in the following year. There can be little doubt he himself commissioned the tomb which Torregiano made for him of terra-cotta in the Rolls Chapel; as it was completed in 1516, the very year he died. As to the altar tomb of Henry VII. and Elizabeth of York, in the Lady Chapel of Westminster Abbey, it was simply a revelation to "those beasts of English," as he contemptuously called them. It is later than the tomb of Young, and was probably finished after Torregiano returned from Italy, where he went in 1518 to enlist workmen; the brass-work of the grate which encloses the tomb being already in place. This grating is purely Gothic, and was made by Nicholas Ewen before Torregiano was brought upon the scene. (See *Gleanings from Westminster Abbey*, p. 80.) The monument of the Lady Anne of Cleves, who died in 1557, is attributed to Haveus, and may be reckoned one of the first of a long series at Westminster in which the Italian taste predominates. Seven years before, in 1550, Brigham had made the monument of Chaucer in the Poets' Corner, and it is completely Gothic in its features. But as far back at least as 1523 the monuments of the second and third Lords Marney, who both died in 1523, were set up at Layer Marney, in Essex, in an advanced Italian style, and of terra-cotta.

If we assume, as we may, if only for the sake of advancing the clearness of a negative conclusion, that the first distinctly Palladian building, as distinguished from Elizabethan, is Caius College at Cambridge, begun probably by Haveus in 1565,—although, as I have said, true Palladian was as yet unknown,—we shall be surprised how much earlier purely Italian monuments appear in churches. We can hardly reckon the terra-cotta tomb at

STEWART MONUMENTS ELY CATHEDRAL.

M

Arundel, but the Dormer monuments at Wing in Bedfordshire date between 1541 and 1552. The monument at Castle Headingham to the fifteenth Earl of Oxford, who died in 1539, though it is not Gothic, can hardly be described as Palladian. But the Audley monument at Saffron Walden, which has many points of resemblance to the tomb of Henry VII., cannot be dated much later than 1544. Several Darcy monuments must have been set up at about the same time in the church of St. Osyth. The date of Sir Anthony Browne's sumptuous tomb in Battle Church in Sussex can be fixed with greater certainty; because it was made in his lifetime to commemorate his wife Alis in the first place. She died in 1540, and Sir Anthony himself in 1548, so that the work must be placed between those years. He had married, secondly, "the Fair Geraldine," Lady Elizabeth Fitzgerald. As this marriage took place in 1543, it seems not unlikely that the tomb was completed by that time. Lady Elizabeth did not take the pains to fill up the date of Sir Anthony's death. At Borley in Essex there are some fine tombs of members of the Waldegrave family, the latest of which, a beautiful arrangement of six composite columns, must be dated before the end of the century. At Chipping Hill, in the same county, there are excellent examples: one of them with a most carefully proportioned tablet which must assuredly have been designed by the architect of the Borley monument. The Rich memorial at Felstead is perhaps the best of those figured by Mr. Chancellor in his splendid volume on Essex *Sepulchral Monuments*. It is dated 1568, but was probably put up a little later. Some of the Stewart monuments in Ely Cathedral are of the same period. For other examples, I must refer the reader to Mr. Chancellor and to local church histories:

but enough have been cited to show that very pure Italian taste was already abroad before the commencement of the first Classical building in England.

Scarcely behind church monuments in assuming a Classical garb were mantelpieces,—in fact, it is only by a careful comparison of the best dated examples that I have arrived at the conclusion that the monuments precede them. Some of these Elizabethan chimneypieces are extremely quaint and picturesque; but the purer Palladian taste and feeling do not come in till a comparatively late period. The fireplace was a favourite object on which to bestow a subtle device, or to display a redundance of heraldic ornament. It was thought appropriate to decorate it with uncouth emblematic figures,—with Flora and Pomona, with bears and lions, with cupids and nymphs,—and this fashion continued long after the full measure of Classical proportion had been attained by the architects. Fine examples of alabaster inlaid with coloured marbles are at Loseley and at Cobham, and others in plain freestone are at Haddon and at South Wraxall. Some magnificent mantelpieces were in London, and a few rescued from old city mansions are at South Kensington; but the destroyer has of late years been very busy among them, and soon there will be few, if any, left. There is one where we should least expect to find it: in one of the new rooms in the western buildings adjoining Westminster Hall. It is said to have come from the Star Chamber. A very fine example is in a bedroom at Knole, and has a pointed arch. At Postlip in Gloucestershire, a fine armorial mantelpiece of the local stone is figured by Nash. One of oak, at Speke in Lancashire, is dated 1598, and there is a beautiful chimneypiece in the Charter House. It would be easy to multiply examples: they occur even in comparatively small

SOUTH WRAXALL.

houses, such as Cheney Court, near Box, Wiltshire, a dower-house of the Spekes of Hazlebury, where the stone chimneypieces rise to the ceiling and are carved with armorial bearings.

Nothing but some very untrustworthy engravings remain of Gresham's Royal Exchange. He undoubtedly brought over Flemish artists to design and build it, and he also probably brought Haveus, the architect, who is already mentioned as having been employed on the tomb of Anne of Cleves. Mr. Papworth speaks of him as "Theodore Haveus, or Heave, of Cleves." To him is attributed, rightly or wrongly, the first distinctly Classical building in England—Caius College at Cambridge. The authorities of that University, some twenty or thirty years ago, seem to have gone on a crusade against what was ancient or interesting in the fabric of their colleges and public buildings. At Pembroke they destroyed the oldest relics in the University, and employed a modern architect to lengthen Wren's chapel—a vandalism of which we shall have more to say. Other atrocities followed; and finally, as a crown to the whole movement, the precious, unique, exquisitely-proportioned little buildings of Caius were handed over to the tender mercies of the same hand which had ruined Pembroke. The new front is described by the guides as "in the French château style," or "that of the French baronial mansions of Francis I.," whatever it may be. Without seeing it, one would say that nothing more inappropriate for the situation could possibly be conceived than a French country château in the centre of Cambridge, in close proximity to two of the greatest architectural ornaments of the University, —King's College chapel and the Senate House,—to say nothing of its interference with the oldest, and in some respects the best Italian buildings in England. But there is little of the French or any other style about the new building. The whole thing must come down, and will, when the College or University authorities return to their senses; but even pulling down will not restore the chapel or the "Gate of Humility."

This gate opened into the College from the street, and was part of a scheme of symbolism very characteristic of the Elizabethan age. By humility the earnest student was to approach his work. It is now in Senate House Passage, or so much of it as was preserved at the removal. We need not describe it, as we have no guarantee that it resembles the original erection, and, judging from analogy and experience, have every reason to suppose it is as much altered as any other.

But the second of these symbolical gates is fairly intact; and, as the very first building in an avowedly Italian style, it is of such transcendent interest in architectural history that its preservation by the present Cambridge arbiters of taste is little short of a miracle. It is, in point of date, the oldest of the three, having been built in 1565; as we read: "On Saturday, the 5th of May, in the year of our Lord 1565, at four in the morning, after offering up prayers to God that our College might enjoy both a prosperous commencement and eventual success, and that all its members might prove men of integrity, lovers of literature, serviceable to the state, and fearing God, we laid the first and sacred stone of the foundation." The date is given wrongly in most books on Cambridge, as Mr. J. W. Clark has proved. It was finished in 1567, and has two inscriptions: one on the eastern side, "Virtutis"; and one on the western, "Jo: Caius posuit Sapientiae." There are two pilasters of the Ionic order, but the archway between is pointed. Above are two storeys with "cottage-headed" Gothic windows, and above them a pediment.

GATE OF HONOUR, CAIUS COLLEGE.

At the south side is a turret set corner-wise and surmounted by a small crocketed stone cupola. The whole composition is extremely picturesque, and it adds to our interest to know that Dr. Caius resided in the rooms over the gate until just before his death, which happened during a visit to London in July 1573. His body was buried in the chapel, on the north wall of which is a beautiful alabaster monument with Corinthian columns and a canopy. The inscription is simply, " Fui Caius," and round the frieze of the canopy are the words " Vivit Post Funera Virtus." This monument, in accordance with what has already been observed, is in a much more advanced style than any of the College buildings. The chapel which it ornamented has been repeatedly altered and "restored"; and the east end, with its Ionic columns and broken pediment, has been replaced by an apse designed by Mr. Waterhouse in his " French château style." The monument of Dr. Caius was removed from his grave and placed high up on the wall near the chancel in 1637.

The justly famous " Gate of Honour " was not built till after the death of Dr. Caius. He is said to have dictated the design to his architect before his death. It is curious to observe in architecture, as in many other arts, that first attempts are often so good. In Egypt, the sculpture of a period so remote that it cannot be dated is not only the best of its kind, but many sculptors and others have acknowledged that a diorite statue of a king—the first royal statue in the world—has never been surpassed. So too, in our own country, some works of the thirteenth century, erected while pointed architecture was in its infancy, remain unapproached. This first effort to build in the new style long stood by itself. There was nothing to compete with it. In the present state of architectural taste it is not likely, in our day at least, to be surpassed. In Ireland, some of the recent buildings in Kildare Street, Dublin, though, like those in Cambridge, small in size, vie with anything built in London since the wanton ruin of Burlington House.

The Gate of Honour is usually dated in 1574-75. It consists of a gateway, a storey of the Corinthian order, with a pediment, and a plain stone cupola, hexagonal in plan, so as to give a very pleasing and varied effect. The whole building is not above 30 feet in height. The lowest storey is, architecturally, the most interesting; for in the pointed doorway we see the efforts of the old Gothic tradition, still prevailing doubtless among the workmen employed, endeavouring to accommodate itself to the Italian views of the architect. Heave, or Haveus, who has been already mentioned, was probably the architect both of this gate and of the others; but it is now impossible to be certain. The Gate of Honour is, of course, dwarfed by the great buildings now surrounding it, but groups well with James Gibbs's Senate House.

The fashion set by Caius at Cambridge was soon followed in other parts of England. Little Shelford in Essex was built for Sir Horatio Pallavicini in or about 1576, and is described as "the first house purely Italian." Some additions to Windsor Castle, now part of the Royal Library, are of this period; and the interior shows purely Italian details. The other examples speedily become too numerous to be mentioned singly; and the names of great architects—Hawthorne, Thorpe, the two Shutes, Smithson, and some who seem to have been simply builders, or, as we should say, contractors, like Warde, Williams, and Hall, who are all mentioned as working in Elizabeth's reign have come down to us. Burghley, Hatfield, Cobham, Holdenby,

Lyveden, Wollaton, Kirby, and many hundreds of smaller houses still testify to the originality of their taste, and to their eye for the picturesque: a quality never forgotten, even when they tried to conform to the Palladian rules as newly interpreted to them. The four books of Architecture were first published at Venice in 1570, and rapidly became known in England. In 1550, John Shute had been sent to study in Italy by John Dudley, the ill-fated Duke of Northumberland; and in 1563 he published his *First and Chief Grounds of Architecture*. It was reprinted in 1579 and in 1584, and must have exercised a great influence on the taste of the time. In a dedication to Queen Elizabeth, he says his work had been approved of by King Edward VI., " whose delectation and pleasure was to see it and such like"; and he goes on to say that having made many sketches,—" trickes and devises " he calls them,—he thinks it well to publish some of them for the profit of other people, and adds, "Wherein I do follow not onelie the writinges of learned men, but also do ground myselfe on my own experience and practise, gathered by the sight of the Monumentes in Italy." There is a volume of Thorpe's drawings, chiefly details, in the Soane Museum. Before the end of Queen Elizabeth's reign, namely, in 1578, Lomazzo's *Trattato dell' arte della Pictura, Scultura, et Architectura*, printed at Milan in 1585, was translated and published in English by Richard Haydocke. This book was dedicated to Sir Thomas Bodley, who about the same time was engaged in rebuilding Duke Humphrey's library at Oxford. There is a frontispiece or engraved title with portrait of " Jo. Paul Lomatius " and his translator Haydocke, who is described as a " Student in Physick." The title runs thus: *A Tracte containing the Artes of curious Paintinge, Carving, and Buildinge*.

There can be no doubt that these publications influenced the mind of a youth who, born in 1573, was destined to do more than any one else to commend the Palladian style of architecture to his countrymen; but before we proceed to examine the career of Inigo Jones, it may be well to pause and answer such simple questions as: What are the elements of Palladian, and how does it chiefly differ from the styles in vogue before it, and especially that one which, for want of a better name, we call Gothic?

There is a strong similarity in all good architecture. The impression to be produced by a building should be threefold. We should be able to see in it harmony of proportion, an expression of stability, and, thirdly, ornament. In other words, a building with architectural pretensions ought to be so proportioned in plan, in elevation, and in parts to the whole, that, without anything else, it should give to the mind of the spectator, through his eyes, such a feeling of pleasure as he derives from a grand or sweet musical composition, or a sublime piece of poetry, or a beautiful painting. This impression is rare, but not transient. It is constantly renewed by the same object, even though that object presents itself differently to different minds. Long before I began to analyse it, I used, at some trouble and expense, to go out of my way to obtain the pleasure of a passing glimpse of Salisbury Cathedral. In poetry, can any one forget the first reading of, say, Mrs. Browning's exquisite lyric, "He giveth His beloved sleep"? There are not many harmonious pictures painted at the present day, but who can see the Ansidei Madonna without pleasure, or look at the delicately balanced colours of the Waterloo Van Eyck without a thrill? When the poet wrote:

> My love is like a red red rose
> That sweetly blows in June :
> My love is like a melody
> That's softly played in tune,

he might well have gone further and taken a third simile from architecture. He might have added that she resembled a Gothic spire or a tapering marble column. There would have been good precedent in the *Song of Solomon*, where the hero is compared to the tower of David, builded for an armoury, and his nose is likened to "the tower of Lebanon which looketh toward Damascus." But so do all the arts, if only they are true, unite with poetry, and display essentially the same qualities.

These things, then, harmony and proportion, which are so necessary in music, painting, and poetry, are still more necessary in architecture. We are told that So-and-so does not admire Gothic architecture, or does not admire Classical and the rest of it. But this is pure rubbish. No cultivated man can admire bad architecture, by what name soever its author may call it. If the man who does not admire Classical architecture sees the Parthenon, he straightway admires it—he cannot help himself.

So, too, we expect in good architecture, Classic or Gothic, or anything else, to see stability. A building should look secure. You should feel when you approach it that it will not fall on your head. Wren said building should be for eternity.

Lastly, a building should be ornamental. By "ornamental" I do not mean "covered with ornament." It is surprising how little applied ornament has to do with beauty. There are warehouses in the City, of recent growth, which are among the most monstrously hideous erections the world has ever seen, and yet are almost built of ornaments. One is in my mind at this moment—a monument of deformity and ugliness. It has no harmony, no look of stability, no real ornament. It has marble columns, but they are of a length so contrary to all the rules, that even their bronze gilt capitals will not redeem them from offensive ugliness. Everything else is of the same character. The architect, who, by the way, has carved his name on a cornerstone, must either have been absolutely ignorant of his art, or else have wished to try if Palladian rules could be violated with impunity. One grudges to see such costly materials wasted. This is only one of a hundred examples which may be counted in a few minutes' walk through the City. Proportion is as necessary to Gothic as to Classic, and the present degradation of architecture is almost wholly attributable to its disuse. A curious example occurs in a book in which we should not look for it. Professor J. Henry Middleton, in his work on *Illuminated Manuscripts*, says: "The sixteenth-century tapestry in the great hall at Hampton Court is a striking example of the way in which gigantic figures may destroy the scale of an interior."

The mediæval architects had very strict rules of proportion at first. In the plain Early English or First Pointed style it was necessary. This is particularly visible in Salisbury Cathedral and may also be seen in the beautiful church of Climping in Sussex, and in the slightly more highly ornamented church of Skelton in Yorkshire. As time went on, architects seem to have thought too much of carving and decoration, with a corresponding degradation of their style; but there is excellent proportion displayed in some Perpendicular churches, as for example Wakefield, Newark, and Coventry, and there are innumerable proofs that the architects studied it carefully.

At the so-called Gothic revival of fifty years ago, proportion fell into disuse, and a disastrous effect has been produced upon design by the idea that detail alone is important. Few architects of the revival escaped it. Hardwick, in his hall of Lincoln's Inn, though it has since been much changed and spoilt by another architect, and his Philological School in the Marylebone Road, proves the need of proportion even in Gothic work. St. Luke's, Chelsea, built by Savage in 1820,—the first revived Gothic church in London,—shows excellent proportions, though poor and even bad in details. But the great safety of the Palladian style lay in the strictness of its rules. What was often done by chance in Gothic was made certain in the style of the Renascence. An architect might not be a genius, but so long as he took care not to transgress the proportions laid down for him, his building could not but avoid any gross error.

The main difference, in England at least, between the old and the new schools, was the attention paid to proportion by the greatest architects, the introduction of columns and pilasters instead of buttresses, and a system of fenestration suited to the improvement in the manufacture of glass. Tracery in windows was discarded as no longer necessary, though it lingered in many places in the shape of those cross-mullioned windows which we see in Brympton House, which is very probably by Inigo Jones. The columns and pilasters had to follow certain patterns, and were of so many diameters according to the order. Cornices, which in the hands of Inigo Jones and Wren became so marvellously ornamental, superseded the parapets, pierced or embattled, of the Gothic architect; and the most beautiful of all the features of the style, the portico, gave us such varieties of charm as we see in St. Paul's, St. Martin's-in-the-Fields, St. George's, Bloomsbury, and many other buildings, religious and secular. The portico, as we see it in Apsley House and many other buildings, was a useless excrescence, meaningless and expensive; but such things belong rather to the "Grecian" period, to which I shall refer later on.

There was another reason why Palladian became popular. It is pre-eminently the style for the Protestant church. No church of the so-called Gothic revival equals in convenience for congregational worship even such a comparatively poor building as St. James's, Piccadilly, while Gibbs's St. Mary le Strand is everything that a church should be. For domestic purposes, it obtained a strong hold; and where no very prominent architectural treatment was required, survived till very lately, local builders where good stone abounded having inherited certain traditions. But these traditions, under the influence of the modern architect, who thinks ornament, and not proportion, is the most important element in architecture, are fast fading away. The anomalous style has come into existence; and some of the most hideous buildings ever laid as burdens upon the earth have been and are being erected. As Garbett said forty years ago, in his *Principles of Design*, of the architect of his day. "He makes a change not for the sake of Truth, but for the sake of change."

The great Italian architect, Palladio, constituted himself the prophet of Vitruvius. It is needful, in order to understand fully the learned architectural style, that we should know who these two remarkable men were. Andrea Palladio, born in 1518, being fond of architecture, as he tells us himself, from his youth, was particularly attracted by the buildings of the ancient Romans. In his opinion, they excelled all who have been since their time in building well. He therefore proposed to himself Vitruvius, the only Roman writer on architecture, as his "master and guide."

His book on *Architecture* was published at Venice in 1570. In that city, always remarkable for its architecture, Sansovino was already practising. As Palladio says in his preface, it was in Venice that " Messer Giacomo Sansovino, a celebrated sculptor and architect, first began to make known the beautiful manner, as is seen (not to mention many other beautiful works of his) in the new Procuratia, which is the richest and most adorned edifice that perhaps has been made since the ancients." The book of Palladio was several times translated and printed in England; and Lord Burlington, having given Isaac Ware, himself a good architect, leave to see and use the original drawings of Palladio in his collection, Ware was induced to have some of them engraved, and to publish a folio volume dedicated to Lord Burlington, and, indeed, revised by that nobleman, which contains the four books of Palladio. The first relates to the five orders; the second describes some houses he had built, with comparisons with those of the Greeks and Latins; the third is concerned with bridges, piazzas, roads, and other works more like engineering than building; and the fourth treats of the ancient Roman remains still extant in Italy. Palladio was born and lived chiefly at Vicenza, where he died in 1580. M. Quatremere de Quincy says of him, that his good taste led him to take the utmost pains with his plans, to adapt his designs to the wants of the time and to moderate means; that he knew how to make a building grand without grand dimensions, and rich without great expense. He adds, that the taste of Palladio found a second home in England, where Inigo Jones, Christopher Wren, James Gibbs, Burlington, Chambers, and many others naturalised his plans, his façades, the happy adjustment of his forms, his profiles, his proportions, and his details. His best work is now to be seen at Vicenza; Quatremere figures his Basilica there, as does Fergusson, as well as his Villa del Capra. To give an idea of his teaching, we may quote from Ware's translation the whole

THIENI PALACE, VICENZA. BY PALLADIO.

of a chapter headed " Of the five orders made use of by the ancients."

" The Tuscan, Dorick, Ionick, Corinthian, and Composite, are the five orders made use of by the ancients. These ought to be so disposed in a building, that the most solid may be placed

undermost, as being the most proper to sustain the weight, and
to give the whole edifice a more firm foundation; therefore the
Dorick must always be placed under the Ionick, the Ionick
under the Corinthian, and the Corinthian under the Composite.
The Tuscan, being a plain, rude order, is therefore very seldom

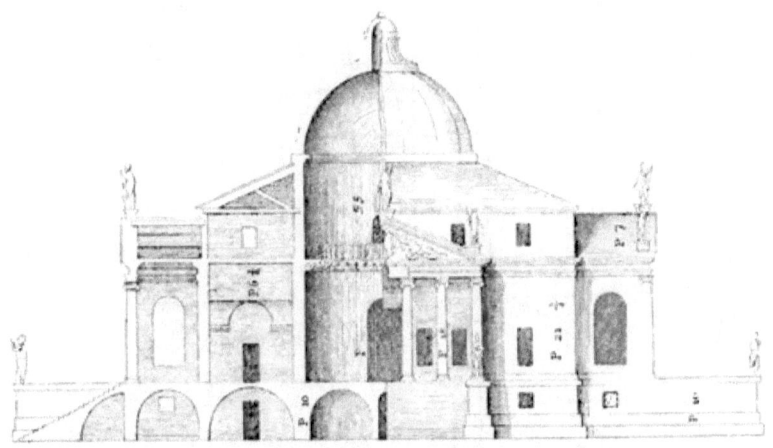

ALMERICO PALACE, VICENZA.   BY PALLADIO.

used above ground, except in villas, where one order only is
employed. In very large buildings, as amphitheatres and such
like, where many orders are required, this, instead of the Dorick,
may be placed under the Ionick. But if you are desirous to
leave out any of these orders,—as, for instance, to place the
Corinthian immediately over the Dorick,—you may, provided you
always observe to place the most strong and solid undermost, for
the reasons above mentioned. The measures and proportions
of each of these orders I shall separately set down; not so much

according to Vitruvius, as to the observations I have made on several ancient edifices. But I shall first mention such particulars as relate to all of them in general."

This chapter from the first book will give a very good idea of the views he endeavoured to impress upon his pupils. Inigo

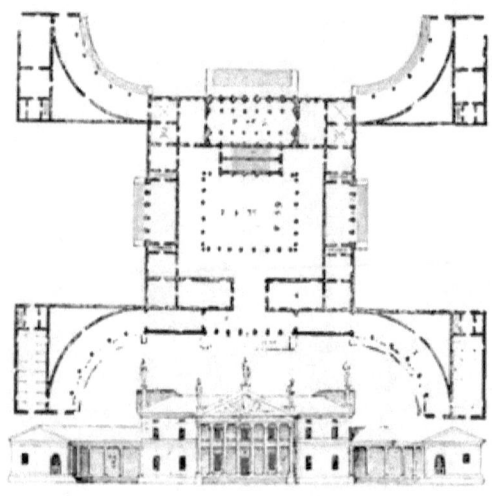

MOCENIGO PALACE, BY PALLADIO.

Jones, we know, had a copy of his book, and no doubt profited by seeing his buildings at Vicenza, Verona, Venice, and other places, as well as those also which he specially praises by Vasari and Sansovino.

Very little is known of Vitruvius, except that he must have lived in the time of Augustus, to whom he dedicated his ten books concerning Architecture—the only Roman work on the subject which has come down to us. From certain allusions he

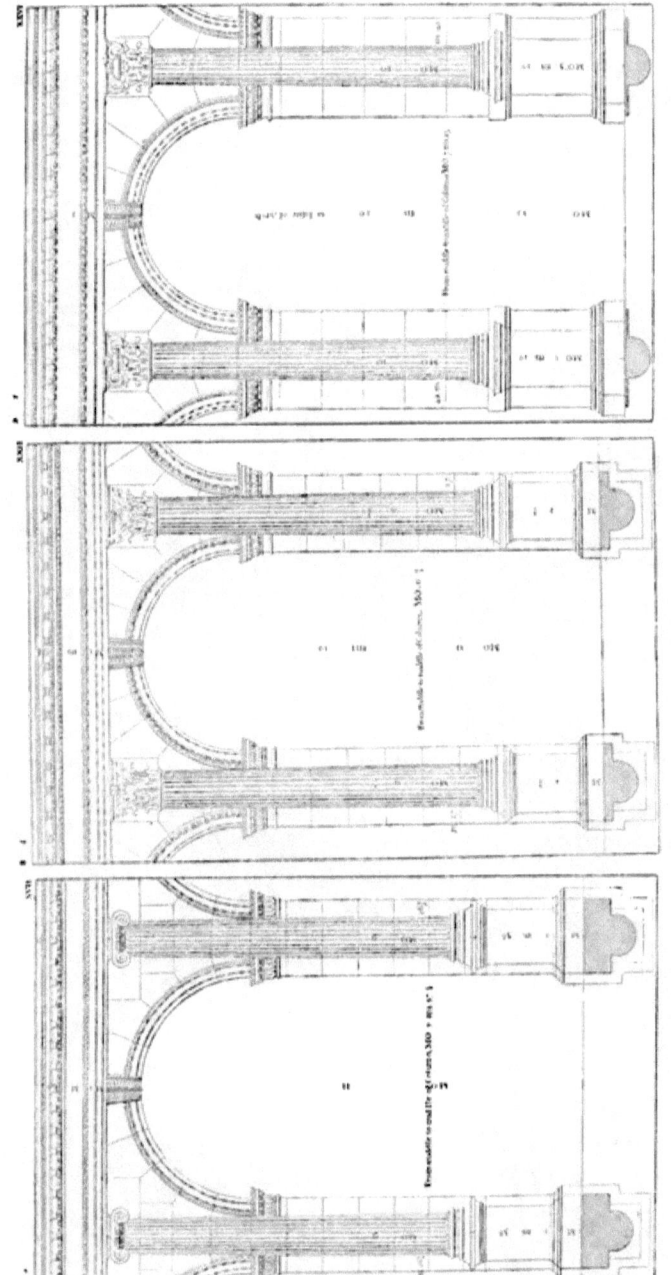

PROPORTIONS OF IONIC, CORINTHIAN, AND COMPOSITE COLUMNS.
(From Ware's *Palladio*.)

was an old man when he wrote, and was, moreover, a man of short stature.

For purposes of reference, it may be worth while to give the principal measurements of the different orders in use among the architects of the school of Palladio.

It should be premised that different architects affected not only different orders, but different proportions: Wren, for example, using one—the Tuscan Doric—in many varieties of proportion.

For the Tuscan, Sir William Chambers laid down the following proportions: "The height of the column is fourteen modules, or seven diameters; that of the whole entablature three modules and a half, which being divided into ten equal parts, three are for the height of the architrave, three for the frieze, and the remaining four for the cornice: the capital is in height one module: the base, including the lower cincture (which is peculiar to the measurement of this order) of the shaft, is also one module, and the shaft, with its upper cincture and astragal, is twelve modules: in interior decoration, the height of the column may be fourteen modules and a half, or even fifteen modules." It was probably this possible variation in the length of the Tuscan column which made it such a favourite with Wren.

The proportions of the Ionic order are stated as follows: The height of the column is eighteen modules, and that of the entablature four and a half, or one quarter the height of the column: if we divide the entablature into ten equal parts, three are for the architrave, three for the frieze, and four for the cornice. There is much variety in different examples of the capital.

The Corinthian order has the same proportions, speaking generally, as the Ionic, but the capital claims an entire diameter. In the Composite style, the column may be as much as twenty modules.

In concluding a long chapter, I cannot do better than quote some expressions made use of by Professor Banister Fletcher in opening the class of Architecture at King's College in 1890. That they should have been necessary, is in itself remarkable : that they should, apparently, have produced no effect, is a melancholy sign of the present state of the art among us :

"A building without proportion is utterly, hopelessly bad. A building, no matter how simple, if in proportion, is good and pleasing. A building in good proportion, and with ornamented construction, is to be desired, and will give pleasure. No amount of ornament, or even ornamented construction, is of any avail in producing a pleasing effect without proportion. Proportion, then, is the very life-blood of Architecture."

V
INIGO JONES

V

INIGO JONES

A List—Parentage and Name—Birth and Baptism—Visits Italy—A Landscape Painter—Proportion—In Denmark—With Prince Henry—A Scene Painter—Surveyor-General—Numerous Drawings—Method of Working—Stage Experience—Arch Row, Lincoln's Inn Fields—Greenwich—Somerset House—York House—Jones and Stone—New Palace of Whitehall—Design for James I.—Design for Charles I.—The Banqueting Hall—A Reredos—Old St. Paul's—St. Paul's, Covent Garden—Ashburnham House—Country Houses—School of Inigo Jones—Death and Burial.

The career of Inigo Jones has already been made the subject of remark in this book, when I had occasion to mention his Gothic work. We have but meagre particulars of his life and training. Under what master he learned to design in so pure a style we cannot tell; but we can judge from his Gothic work that the eye for proportion, in which he excelled all his contemporaries and most of his successors, was both born in him and also sedulously cultivated. We have reason, moreover, to believe that he was a good draughtsman, though it is by no means certain that all the drawings in the Devonshire and other collections are actually by his hand. Of his buildings, few remain, and still fewer are intact. He certainly made some designs for Wilton, and a bridge there is undoubtedly his. At Widcombe, close to Bath, there is a small, but beautiful manor-house always locally ascribed to him; but it is apparently later. I have mentioned his Gothic work at Oxford already; but he does not appear to have designed anything there in his most characteristic style, unless we assign to him the porch of St. Mary's Church. In London, we had many pieces well worthy of his skill; but year by year they perish. The excellent Society for Photographing Relics of Old London has preserved views of several which have lately been removed. Mr. Alfred Marks enumerates as his: Shaftesbury House, Aldersgate, now destroyed; Ashburnham House, in Little Dean's Yard, Westminster; the Banqueting House at Whitehall; the chapel of Lincoln's Inn, which, as their chief treasure, the benchers handed over for "restoration" to an amateur who had already characterised Inigo's work as "horrible"; the Water Gate of York House, in Buckingham Street; and Lindsey House, with possibly one or two other fronts in Lincoln's Inn Fields. There are two houses very like his work in Great Queen Street. Two others, locally ascribed to him, were in Great St. Helen's, and were pulled down in 1892. The portico of St. Paul's, Covent Garden, which has been burnt and rebuilt and finally handed over to a "Gothic" architect for improvement and the side entrances pulled down, with some reminiscences rather than remains in the adjacent buildings of the Piazza, may safely be assigned to him. The staircase of a house on the south side of Chandos Street, pulled down last year, was very likely his, and was very handsome both in design and execution. He lived, we know, for a time very near the spot, before the existing streets were laid out. Lastly, we may trace his hand at Greenwich, where Wren's noble design was specially made to include, preserve, and complete what Jones had begun.

It has frequently been asserted that his father was a tailor near St. Paul's, and had dealings with Spanish merchants, one

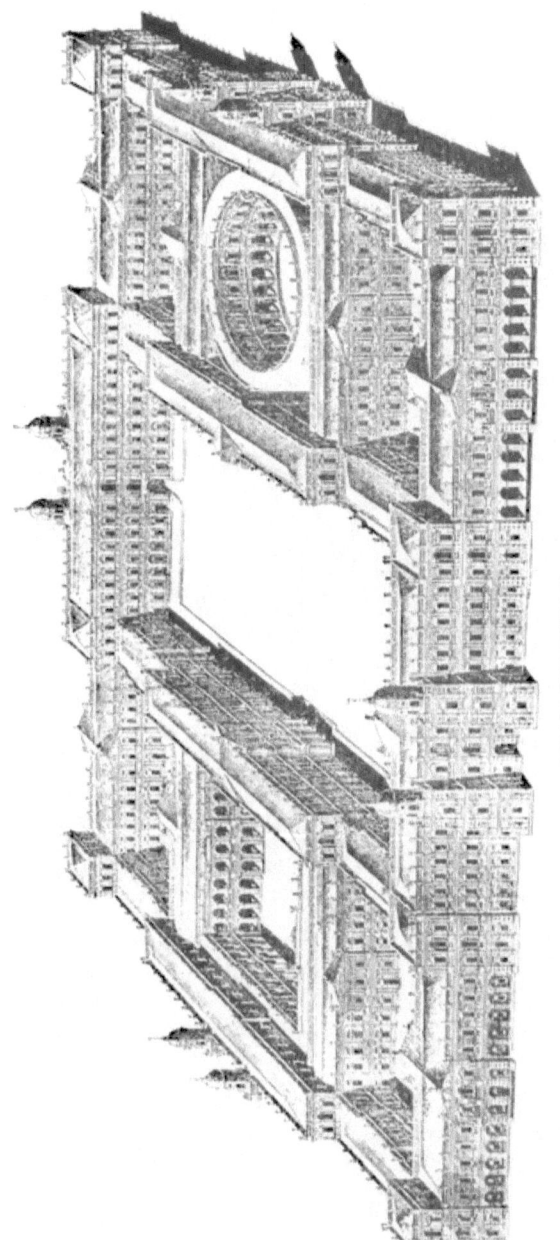

PALACE OF WHITEHALL.
As designed by Inigo Jones, 1619. From an engraving by T. M. Müller, 1749.

of whom gave his Christian name to the son. This story, started after Inigo's death by his friend Webb, and greatly enlarged and improved by Quatremere de Quincy and other imaginative writers, has apparently no foundation. The elder Jones was not a tailor. He lived in the parish of St. Bartholomew's. He was a Welshman by birth, and himself bore the name of Inigo, which he gave to his son. He was a cloth-worker, and had to compound with his creditors in 1589. As to the name, it resembles Jenico, which is not uncommon in Ireland; and both may be diminutives of Ignatius. "Inigo" certainly occurs in Spain; but it does so also in Wales at the present day.

Inigo Jones, the architect, was born 15th July 1573, and baptized fourteen days later in the church of St. Bartholomew the Less: the parish of which his father, a Roman Catholic, was an inhabitant. Inigo, the elder, was not fortunate in business, and his son, the younger Inigo, was apprenticed to a joiner. By some means, with which we are not acquainted, he contrived or deserved to attract the attention of that popular young nobleman, William, third Earl of Pembroke of his family. Pembroke sent him to Italy to study landscape painting, so far as we can judge; and his "travelling scholarship" was continued by another great noble who wished to be a patron of the arts, Thomas Howard, Earl of Arundel, by whose help, apparently, he was able to visit Venice. But we know wonderfully little about his early life and career. He painted a few landscapes, which are still extant, and, no doubt, many more which have perished; and his art seems to have been of that picturesque or scenic character which is so little in vogue at the present day— the art of Claude and Poussin. He wanted to obtain the utmost possible amount of effect with the least possible amount of cost.

and set himself to find out how to attain such an object. This
was a study closely akin to architecture. It has little—too little—
connection with either the art or the architecture most in fashion
now. The coloured photographs which crowd the walls of the
Academy at the present day, would have had no more charm for
him than for his great contemporaries—Claude, or Shakespeare,
or Bacon. He learned, we know not where, without great
expense or great trouble, how a cloud-capt tower could be
taught to rise as the background of a play; or, how a fairy landscape, far reaching by sunny rivers and high-walled cities, could
form a permanent scene. Such devices were unknown in England
at that day. He learnt them in Italy, and he learnt more. He
saw from his experience of people not half so rich, not half so
poetical, not half so noble as the contemporaries and countrymen of Shakespeare, that in England his art would be appreciated properly: the art which showed them grander and more
gorgeously coloured and more artificially designed scenes than
any they could find even in their own beautiful island. But in
order to produce these effects, a knowledge of proportion came
first. Without it, he could have done nothing. With it, he
made a little piece of stage scenery, some thirty or forty feet
square, look a hundred feet—nay, a thousand. If he could have
walked through some of our modern streets and seen with what
success great theatres and international institutions are made to
look no larger than mud huts or thatched cottages, he would
have wondered what our artists had been at for three hundred
years. We should have had to take him to Dublin or to Liverpool to prove to him that scenic effect is still cultivated and
studied, though not in those places which think themselves the
centres of civilisation. The fact is, and it is a fact not to be

overlooked, that the genius of Inigo Jones, like the genius of Wren and Newton, of Mulready and Turner, of Watt and Herschel, consisted in the power of concentrating the mind upon a problem till it was solved, of taking infinite pains until the desired end should be attained. Too many of our modern artists and architects forget this fact. They will not "take thought." And why should they? The general public do not ask for it. The artist's and the architect's employers do not require it. All such things as harmony of colour in painting and of proportion in architecture have been thrown aside. No doubt, they are troublesome things to study; and if the man who employs the artist is ignorant of them, why should time be wasted over them? But we too often omit one important point. Let a man be ever so ignorant about proportion, yet when it is presented to him he recognises its superiority, just as a man who has never heard good music can admire a sweet air the first time he hears it. Without proportion, Jones could not have influenced the taste of his time as he did. But, undoubtedly, there existed a few men, and they chanced to have wealth and influence, who knew good art when they saw it, and accepted it when Inigo offered it to them. His contemporaries are loud in his praise; and many buildings are attributed to him, even in foreign countries, that he can never have been concerned with. His journey to Denmark assigns to him the castle of Fredensborg and the Bourse of Copenhagen; and his possible visit to Scotland, as a member of the suite of the Princess Anne of Denmark, makes him the designer of Heriot's Hospital.

As a fact, however, we do not know of his designing at this period anything more important than the scenery of masques. In 1605, he was thus employed at Oxford. In 1610, he was a kind of stage-manager for the Queen and Court at Whitehall, and had his memorable quarrel with Ben Jonson. About this same time he was appointed Surveyor to Henry, Prince of Wales. On the prince's death, he returned to Italy, where the Italians seem immediately to have appreciated his genius; and two buildings at Leghorn are pointed out as his. Some are also ascribed to him at Venice, but all on slender grounds.

Returning to England, he became Surveyor-General in 1615, and from this time devoted himself to architecture.

Among his first works for the Government, was an alteration of the Star Chamber at Westminster; and he made a complete design for a new building. The drawings are now at Worcester College, Oxford, but were never carried out. It is worth while here to remark upon the large number extant of the drawings of Inigo Jones; and, moreover, upon the fact that of them all, hardly any represent works accomplished. The Duke of Devonshire inherited a great many from Lord Burlington. Others are in the British Museum, and in the Royal Library at Windsor Castle. But of them all, not one seems to have been carried out. It is true, the Banqueting House at Whitehall, and Ashburnham House at Westminster, are very nearly what he left in drawings, but they are not exactly so; and it has been suggested that his drawings were used up and worn out by the workmen employed in carrying them out. I have already, following the lead of Mr. Gotch, endeavoured to show that, among the predecessors of Inigo Jones, working drawings of the modern kind were not in use. Workmen had their traditional rules, and followed them according to the department on which they were employed. But in buildings like those of Jones, where the most delicate proportion was to be the chief feature, and mere

LINCOLN'S INN FIELDS

ornamental workmanship a trifling detail, drawings with careful measurements were necessary. Before his time, architectural drawings were practically unknown. We have no drawings of the great churches of the middle ages. A few occur in France and Germany; and their rarity does but prove the rule. Since the time of Inigo Jones, they have abounded in England. This leads us by inference to a further fact. The carvers, the lead-workers, and the glaziers of King's College Chapel or of Middle Temple Hall set the patterns of the works they executed themselves. They wrought according to the traditions of their forefathers and predecessors. But Inigo Jones was anxious that everything in a house which he designed should be his. He could allow nothing of a "rule of thumb" kind. Everything, down to the balusters of the staircase and up to the plaster-work of the ceiling, must be drawn to scale, and, what was more, must be executed by the workmen according to the drawings. In the seventeenth century, as two hundred and fifty years later, we may be sure that the British working man, while professing the most advanced ideas in politics and in all commercial matters except that which concerned himself, was as averse to any mechanical innovation as if he had lived in the reign of Queen Victoria and been called a Radical. But here Inigo's previous career enabled him to carry out and maintain a change which has ever since prevailed. He had been obliged, in designing the scenery of stage plays and masques, to insist on the closest accuracy on the part of the workmen he employed. Any one, who has ever endeavoured to make the background of a theatrical piece, will bear me out when I say that an error of a couple of inches may mar the whole illusion. This Inigo Jones knew well. His men were accustomed to do implicitly what he bid them. His working drawings superseded the old arrangements. It was no longer a case of surveyor and workmen, each skilled in his own department; but it was a case of architect, and architect alone. In this way Jones led up to Wren. After Jones's time, drawings were the rule. The result was twofold. Where an architect was an artist, he might feel sure his work was adequately rendered. Where he was—well, I will not mention the men who are in my mind, but say—a builder, he no longer had the skilled artisans of the time of Queen Elizabeth to make things right for him. A house or a church, however poorly designed, which had good carving, lead-work, masonry, and mouldings, was still tolerable. But under the new system, which I think is to be attributed to Inigo Jones, a bad architect could make good workmen perpetrate bad things; and, on the other hand, Wren and some of his school, who could not possibly have done what they did under the old system, were able to design and carry out buildings which have ever since been a joy to us all. We must put the gain against the loss; but it is needful to remember that Inigo Jones not only revolutionised our old architecture, but that he revolutionised our old workmen, and that Wren and his school would have been impossible but for him. The old system rapidly died out. I have mentioned buildings at Oxford which continued to be made in the ancient fashion; but by degrees working drawings superseded every other method, and Wren could count on the closest fidelity to his designs. In one particular, as we shall see, he reverted for a reason to the old way; and when he found a great artist like Grinling Gibbons, he left him at liberty to follow the bent of his inclinations and his genius. As a result, we have the glorious series of carvings which so greatly adorn St. Paul's.

In 1615, Inigo Jones took up the office of Surveyor, and had a house in Scotland Yard. He seems to have been very active in the practice of his profession for many employers besides the Government. Of his London work, but little remains. The western side of Lincoln's Inn Fields was certainly his; but only a few fragments of "Arch Row," sadly mutilated and plastered over, can be identified by the fleur-de-lis and rose he placed on them in honour, it is said, of the King and Queen. His Lindsey House, in the same row, is better visible. The same Ionic order is employed; but here he ornamented it with wreaths on each capital. The house is now divided, and the beautiful centre window is built up. In 1617, he commenced to work at Greenwich, where he built the Queen's House, which now forms the centre of the Royal Naval Schools. Some parts of the Hospital buildings are also of his design, but were completed after his death by Webb. In the Strand, Inigo Jones built the Queen's palace of Somerset House, of which several views are extant. Chambers appears to have followed them in designing the Strand front of the present building. He also built York House, on the site now covered by Buckingham Street, for the Duke of Buckingham, and designed the beautiful Water Gate, carved with the Villiers arms, which now stands in a kind of pit at the foot of the street, and is best seen from the gardens of the Embankment. This gate has often been attributed to Jones's friend Stone, the best sculptor of the day, who is known by other fine works in London. But there can be no doubt that, though the carving is by Stone, the design is by Jones. Many other houses, great and small, have been attributed to Inigo; but the great work of his life is to be found in the two sets of drawings he left for a new royal palace at Whitehall, made, the first for James I., in and after 1619, the second for Charles I., twenty years later.

Three sets of prints from these designs are well known and not uncommon, and will be referred to here; but there are separate drawings, as well, in the collections which have not been published. There are discrepancies in the various sets of prints; and they also differ from the single view of the Banqueting Hall in the first volume of the *Vitruvius Britannicus*. Nor is this all: there are some important differences between all the prints and the actual elevation as accurately set forth in photographs.

Of engravings in which the Hall appears as part of a great design, we may compare those of Kent and of Campbell's *Vitruvius* (vol. ii.). There is no letterpress to the Kent view, except to say that this particular print shows "the front of one side of the Palace, within the Great Court, and section of the buildings at each end of it, with the side of the towers." We see buildings repeated within this court which closely resemble the Hall, except that there is a doorway in the centre on the ground storey of each. By the plan, we learn that the resembling buildings were different internally, and that the farthest from Charing Cross, the more southern, was broken up into apartments. But on further examination, we discover that, while the northern building answers to the Banqueting Hall that now is, there was to be a corresponding building exactly opposite on the other side of the same Grand Court; and this was to be the Chapel. The existing Hall was thus to be only one of four minor features of a court of altogether 919 feet 3 inches in length, the internal measurement being 740 feet; and the great features were to be two elevations facing each other on the right and left, containing grand state apartments, as fine again as the Hall

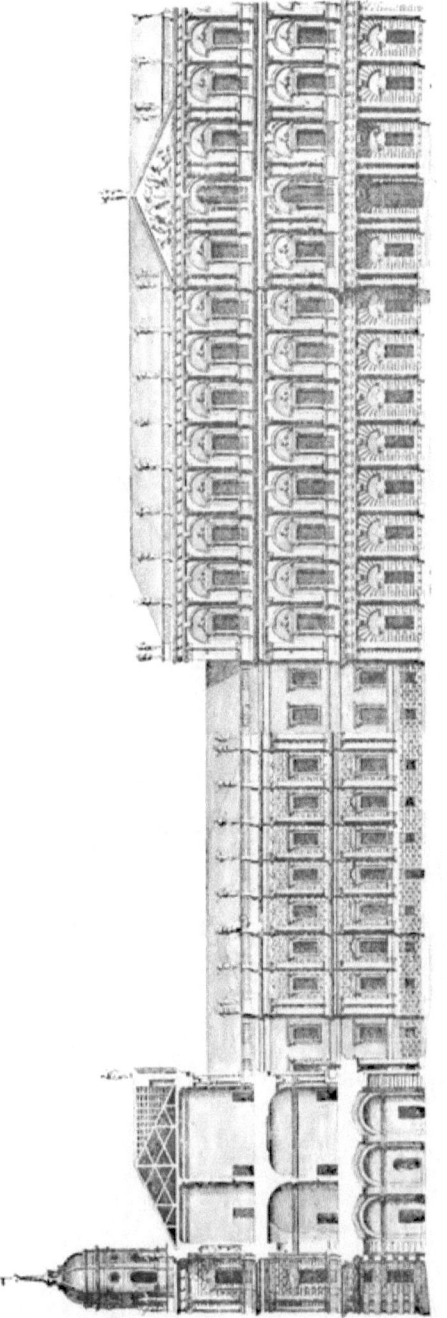

PART OF COURT, SHOWING BANQUETING HALL, WHITEHALL
By Inigo Jones, 1620.

which alone was built. The rest of the first volume of Kent's *Inigo Jones* is taken up with the details of this grand scheme.

CENTRAL PORTION OF UPPER STORY, BANQUETING HALL, WHITEHALL.

It is summarised thus in Fergusson's *Modern Architecture*: "It was proposed that the palace should have a façade facing the river, 847 feet in extent, and a corresponding one facing the park, of the same dimensions. These were to be joined by a grand façade facing Charing Cross, 1152 feet from angle to angle, with a similar one facing Westminster. The great court of the palace, 378 wide by twice that number of feet in length, occupied the position of the street (120 feet wide) now existing between the Banqueting House and the Horse Guards. Between this and the river, there were three square courts, and on the side towards the park a circular court in the centre, with two square ones on either hand. The greater part of the building was intended to be three storeys in height, each storey measuring, on an average, about 30 feet, and the whole block, with podium and balustrade, about 100 feet. The rest, like the Banqueting House, was to have been of two storeys, and 78 feet high " (p. 257).

When we remember that this palace was not only to be constructed of the best and most costly materials, and was to be finished by the greatest artists,—Rubens himself being actually employed to decorate the ceiling of the Banqueting House with a picture of the gods receiving James I. to heaven,—but was also to equal Versailles in size, we need not wonder that the treasury of England failed to sustain the burden. It was to be at least twice as large as our new Houses of Parliament; and as the actual residence of a reigning king, would, of necessity, have been far more magnificent. We get a good idea of what was in Jones's mind from a print of large size showing the projected palace in a bird's-eye view. It was published in 1749 by T. M. Müller, an engraver, together with a series showing the sides. A month's study of them would teach a modern architectural student more than a year in a "mock Gothic" office. The whole thing is, of course, so far as substantiality goes, only a scene-painter's day-dream, a pictorial poem, but well worthy of examination for the improvement of the taste, the information of the fancy, the exercise of criticism,—for it is not without faults,—but above all for an introduction to scientific proportion, the harmony of architecture. Too often the modern architect knows so little about it himself that he cannot teach it to a pupil. As I have said, there are faults in these designs, and one of the worst of them was the difficulty of making a sufficiently wide archway for the old street and right of way of Whitehall; but this might no doubt have been overcome by a man of Inigo's fertility of resource and invention. To me, the difficulty is the unreality induced by impossibility; but here again, though the whole of his delightful tale could not come true, each single incident was beautiful and poetical in itself, and the only fragment which we have has been a source of pleasure to men of taste for centuries.

The next engravings in which we find the Hall represented as part of a great design are of a very different character. The prints so far described relate to the design made for James I. in or soon after 1619. In Campbell's *Vitruvius Britannicus* we have a design thus marked on the plate: "The Elevation of a Design for the Palace at Whitehall towards the Park, as it was presented to His Majesty King Charles I. by the famous Inigo Jones, Anno 1639." Fergusson characterises this as the work of Jones's decline, which it is not, and as "the impoverished makeshift" forced from him by the troubles of the times, which it may be. But it has points of great beauty and shows more experience and more of that power of designing without excessive ornament which he had first displayed at Covent Garden. This second of the two designs, that for Charles, had only two of the four Banqueting Hall elevations, and came but just up to the edge of the street of Whitehall. There are difficulties in reconciling the letterpress with the engravings; but the difficulty about the narrow archway through which the roadway from Charing Cross was to run to Westminster, of which Fergusson says so much, was got rid of by abolishing all that was to stand west of that line, by making the Banqueting Hall and its companion, a chapel, open to the street, and by constructing midway between them a very handsome gateway two storeys high into the principal court. The side toward the river was to be plain but handsome, and very picturesque; the design of the corner towers and the cupolas being as good as anything in the Jacobean drawings. The engraver has neglected to put the points of the compass on his plate, which is calculated to puzzle the reader, and evidently has puzzled Fergusson and others; but assuming that the Banqueting Hall was the building in the print to the left of the gateway, it will be evident that the whole palace was intended to cover only about half the space of the palace originally projected by James I. It was to be two storeys high throughout above the basement.

The Hall is 110 feet long by 55 feet, and the same in height. It is thus a double cube—a very perfect proportion in itself, and one which Barry judiciously adopted for the principal chambers of his Palace of Westminster. The back and front are alike, and show two storeys above a massive but not rusticated basement. The first floor is Ionic, the order being beautifully designed and carved in his best manner by Nicholas Stone, who acted as master mason. The upper storey is Composite, and there is a carved wreath, forming a frieze between the columns, of which four in the centre are engaged, three at each end being flat pilasters. The windows in the lower range are alternately round and pedimented, those of the upper storey being flat.

PORTICO, OLD ST. PAUL'S. BY INIGO JONES.

The parapet is finished with a balustrade, and was probably intended to have been decorated with statues. About 1724 the Hall was first fitted up for a chapel; the original chapel of the palace having been on the river face. For this older chapel Jones had designed a beautiful altar-piece or reredos, which was given eventually to Westminster Abbey, where it was destroyed some years ago by way of "restoration," and the present unsightly and poverty-stricken erection substituted. Jones did not, of course, fit up Whitehall for a chapel; and the chief decorations, except Rubens's ceiling, were, no doubt, by Sir John Soane, who carried out extensive works in 1829.

The repair of old St Paul's was carried on by Inigo Jones during a series of years. Accepting the Norman nave as a building of Romanesque character, he added to it a magnificent western portico, which must assuredly have been the most beautiful of its kind in England. It consisted of ten monolithic Corinthian columns in front, the corner columns being square, and stood out well, with three intercolumniations at the sides. In the course of his work on the Cathedral, Inigo Jones pulled down a portion of the church of St. Gregory, which stood in way of his portico; and his action was bitterly complained of by the parishioners. The portico was completed in the same year, 1631, as the church of St Katharine Cree; and if Jones designed both, as seems very probable, the proof afforded of his versatility is remarkable. The same year also saw the building of St Paul's in Covent Garden: a very curious church in several respects. Like Whitehall, it has been severely restored; but its salient points are still visible. There is an account in Vitruvius of a style in vogue in Etruria; but no example was known to exist. In the British Museum, among some Roman marbles, there is a bas-relief which shows a house in what must be the style Vitruvius describes as "Tuscan." From the description of Vitruvius, Jones made out certain characteristics, such as the massive columns and the overhanging eaves. He had orders from Lord Bedford to make a very plain building,—a barn, in fact,—and is said to have replied, "You shall have the handsomest barn in England." Walpole tells this rather idle tale. There is no reason to doubt that Jones designed his church to harmonise with the piazzas and other buildings of the square upon which it looked. As the square is to eastward of the church, the portico is at the east end. The architect, owing to the custom peculiar to England of having the chancel at that end, was not permitted to make his entrance there. The portico, in short, is a mere adjunct of the square, and no integral part of the church. Though Inigo's hand is very plainly marked upon the whole design, it is well to observe that St. Paul's has been subjected to repeated alterations. It was almost rebuilt in 1688. Lord Burlington restored the portico in 1727. The whole fabric was renewed in 1788, by Hardwick. Finally, it was burnt down in 1795, and rebuilt by the same architect. A modern architect was employed on it in 1872, and carried out extensive alterations; and, in 1888, the west end turret was taken down, and the red brick of the walls was refaced and coloured. A hideous cast-iron railing at each side replaces the original walls and arches.

The building has been alternately praised and abused. One architect, Ralph, considered it the most perfect piece of architecture that the art of man can produce. Horace Walpole thought it wanting in dignity or beauty. Brayley, a good and impartial judge, says it forms a striking object viewed from the market-place, and it is known to have excited Lord Burlington's deepest admiration.

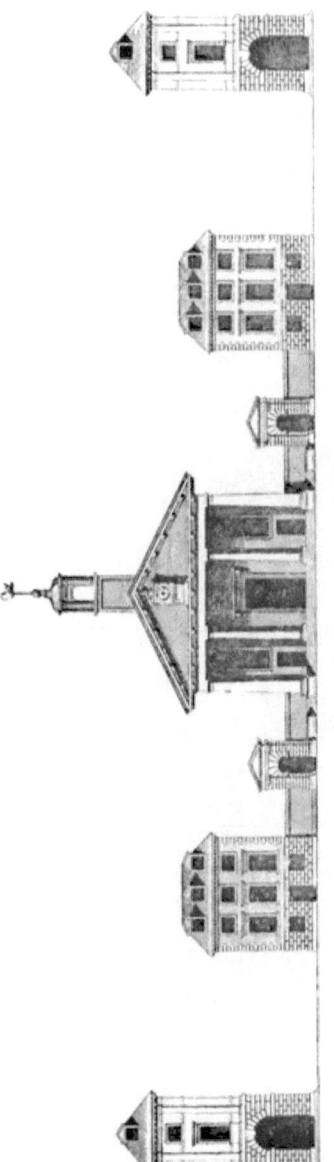
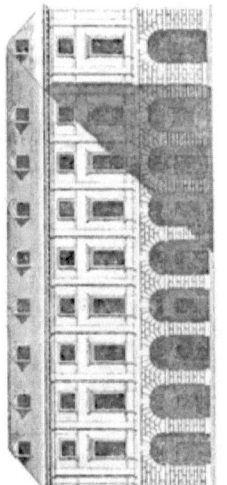
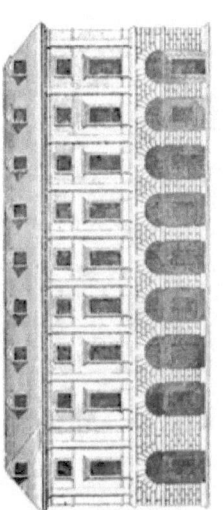

COVENT GARDEN: CHURCH AND PIAZZA

The following more technical description is mainly taken from Brayley. The portico on the eastern front consists of two lofty columns and two square piers of similar character, supporting an angular pediment. The pillars diminish considerably towards the capitals. The interior is very plain; the chief entrance being from the west end. The ceiling is flat. The Tuscan style allowed the frieze to be dispensed with, and other elements of cost to be omitted; while the projection of the roof enabled the architect to obtain considerable picturesqueness, as well as dignity. The roof, covered with slate, was formerly of tiles, which must have added greatly to the effect.

Fergusson, whose praise of this building is not very warm, adds sensibly: "No one can mistake its being a church; and it would be extremely difficult, if possible, to quote another in which so grand an effect is produced by such simple means."

There is no doubt as to Inigo's own opinion. He desired, by his will, that on his monument in the church of St. Benet, Paul's Wharf, should be placed views in relief of the portico of St. Paul's and of the church of Covent Garden.

Ashburnham House stands within Little Dean's Yard at Westminster. Its exterior is of brick, unrelieved by ornamental features. At one time, as in all Jones's work, the proportions were sufficient to call attention to it; but the addition of a storey, and other changes, deprived it even of this characteristic. The last canon who lived in it seems to have done his best to preserve what was left. It now belongs to the school, and is taken excellent care of, and freely shown. The late Sir Gilbert Scott is said to have doubted that it was designed by Jones; in other words, he wanted to pull it down, but additional proofs have come to hand. Nevertheless, it is right to mention that Batty

Langley, in 1737, says it was built by Webb from Jones's designs. To this opinion Mr. Marks also seems to incline.

The hall is not remarkable, being merely a well-proportioned rectangular apartment. The staircase, which opens from it on the eastern side, is the principal feature of the design. The house, it is clear, was built on part of the site of the Refectory of Westminster Abbey; and a thick wall of mediæval masonry divides it longitudinally. Remains of something like a buttery hatch still remain on the ground-floor. The staircase is described in precise language in Britton and Pugin's *Edifices* (ii. 90). "Of nearly a square shape, with four ranges of steps, placed at right angles one with the other, and as many landings, it was the passage from the ground to the first floor. Its sides are panelled against the wall, and guarded by a rising balustrade: the whole is crowned by an oval dome, springing from a bold and enriched entablature supported by a series of twelve columns. At the landing are fluted Ionic columns." The uppermost landing gives access to a dining-room by a very deep doorway cut through the refectory wall. The alcove in the dining-room is by a later hand. Another doorway admits to the anteroom; and that, by a beautiful doorway, to the drawing-room, in which there is a richly ornamented ceiling by Inigo, formerly surmounted by a small oval dome or lantern, removed no doubt when the upper storey was added to the building. Views of Ashburnham House, within and without, appear in many books, such as Ware, Batty Langley, Smith's *Westminster*, Britton and Pugin, and others. The staircase has been many times imitated; and certainly no better model can be conceived. It has been introduced also into pictures. Mr. Laslett Pott makes it the scene of his *Disinherited*. Sir John Soane had large drawings

COBHAM HALL, KENT

T

made of it for his lectures at the Royal Academy. They are now with the lectures in the Soane Museum. There are five views of Ashburnham House in the series published by the Society for Photographing Relics of Old London.

A house in St. Martin's Lane, No. 31, is said to be by Jones, but is not very interesting. In 1634, he designed a monument for St. Giles's-in-the-Fields to commemorate his friend, George Chapman the poet. He is also often credited with the stables of Kensington Palace, then Nottingham House; but it is pretty certain that for "Kensington" we should read Kennington, where Prince Henry had a house. Charlton House, near Greenwich, is also attributed to him, with the Fellows' Lodgings, Christ's College, Cambridge, and with Brympton, a manor-house in Somerset. A portion of Brympton is very like his work.

In the country, Jones's best piece of work is a little Ionic bridge at Wilton, exactly imitated by Wood at Prior Park, near Bath. He carried out some work at Wilton House, but not very much; and both there and at Amesbury and Greenwich, buildings from his design were executed after his death by Webb. The central block of Cobham Hall, near Gravesend, containing the beautiful Music Room, is always, though without absolute proof, attributed to him. It groups charmingly with the older work, with its engaged Corinthian columns, its cornice and its doorway. At Cambridge, the Pepysian Library at Magdalen has been very doubtfully assigned to him; and at Oxford, the beautiful porch of St. Mary's Church. A porch at Magdalen used to be called his. It was destroyed some years ago, and a so-called Gothic design by Pugin substituted. It is very ugly and unsuitable. There is a gateway in the Botanic Gardens, opposite Magdalen College, which resembles his work.

Two or three country houses are, or have been, very like his work. Of Coleshill in Berkshire we cannot be sure, though it is positively asserted to have been built by Jones in 1650, and certainly looks very like his handiwork. Lord Burlington believed in Coleshill, and employed Ware to make drawings of it (see *Vit. Brit.* v. 86). A gateway at Holland House is by Jones, having been carved by Stone, and another gateway, removed from Beaufort House, Chelsea, was presented by Sir Hans Sloane to Lord Burlington, and is now at Chiswick.

The best authorities as to the life and work of this great artist are not to be found in the most popular books. Peter Cunningham wrote a biography which is esteemed the best, but it is rather scarce; and as Cunningham, though he knew London well, was no architectural critic, we are sometimes at a loss. The best account of Inigo Jones is undoubtedly that in the studiously dry article, written anonymously, in the *Dictionary of National Biography*. It is so crammed with facts and dates that it literally took me three days' hard work "to make it up." It is almost without criticism, and the reader can therefore learn only half, and that the least important half, of Jones's strange story from it. Nevertheless, two very different, though intimately associated, kinds of readers are interested in it. We turn to it to learn about the history of London; and we turn to it also to help us to a clear understanding of the chief architectural problems of that day. The Civil War prevented Inigo Jones from founding what would now be called a school, but he left traditions; and his loyal friend and executor, John Webb, took care, when settled times came again, that the great teacher's name should not be forgotten. With a rare generosity, while he constantly used the designs left to him that they might be used,

BRYMPTON.

he as constantly attributed them to "the vanished hand," and did not even claim some designs in which his own share must have been by far the largest. The Civil War put a stop to all artistic development, and we have nothing like the brilliant following which Wren left: no Hawksmoor, Gibbs, Kent, or Burlington. A very good and appreciative account of the works of Jones appeared in the *Portfolio* in 1888, from the pen of Mr. Reginald Blomfield.

His popularity and fame as a great architect are curiously attested by the existence, as late as the middle of the eighteenth century, of a house in the Strand called after him. Dart's great work, the *Cathedral of Canterbury*, was published in 1726 by "J. Smith, at Inego Jones's Head, near the Fountain Tavern."

His last years fell on troubled times. He followed Charles to Oxford at the outbreak of the war, and is said to have lent the King a sum of money. We next find him shut up in Basing House while it was besieged by Cromwell. He was able to prove that he was not there as a belligerent, and eventually got off with a fine of £545 and a payment of £500 to compound for his estate.

On the 21st June 1652, having been in declining health for some time, he died at Somerset House in the Strand. He had never married, and the bulk of his property went to cousins, one of whom had married John Webb, who was a good architect himself and had much assisted Inigo in his later years. He inherited all his drawings and designs with an express idea that they should be kept together, as they were indeed at first; but William Webb, John's son, seems to have been careless of them. Clarke's collection of the drawings was bought from William's widow, and was left in 1730 to Worcester College. Kent and his friend Lord Burlington also formed a large collection, which has lately been removed, I hear, to Chatsworth; but the Duke of Devonshire exhibited a great many at the Burlington Club in 1884, and some more at the Royal Institute of Architects in 1892.

From Somerset House, Jones's remains were removed to St. Benet's, Paul's Wharf, where they were buried with those of his father, the Welshman. By a curious coincidence, St. Benet's was selected a few years ago for the use of a Welsh congregation, who opened it for their service in 1867. But the actual church in which Inigo was buried was destroyed in the Great Fire; and the present one was built by Wren in 1683. It is in red brick, and very picturesque, with carved festoons over the windows. Inigo left £100 for his monument, as well as £10 to the poor, and £100 for his funeral expenses. The monument, in marble, was of course destroyed with the church; but the burial is recorded in the register, 26th June 1652.

In concluding this chapter, I must refer the reader to some expressions made use of by Mr. Blomfield in the articles already referred to as having appeared in the *Portfolio* in 1888. I quote them because Mr. Blomfield is a practical architect, which I am not; and I wish to show that what I have said as to the chief merit of Inigo Jones, his sense of proportion, and as to its neglect at the present day, is not too strong. Mr. Blomfield observes that " in all his studies, the one point on which he concentrated his energies was proportion." He resolved designs into their constituent parts. He showed, for instance, that the Temple of Jupiter " was based on a series of circles, and its proportions arrived at by dividing the largest diameter into six parts and variously recombining the parts. No man saw more clearly that proportion is the keystone of architecture." Mr. Blomfield

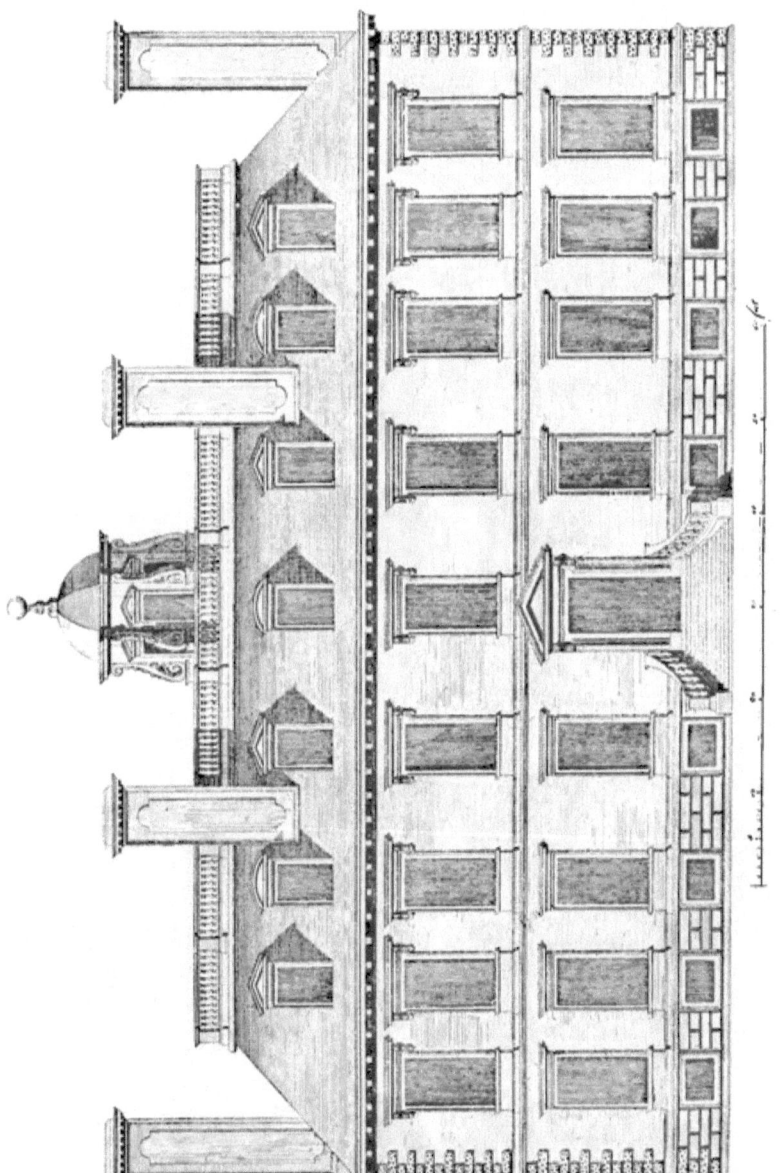

COLESHILL.

asserts truly that what "values" are to the painter, proportions are to the architect; and he follows with a passage which I am glad he has written and not I, though I agree with every word of it, and could adopt it as the motto or argument of this whole volume. "The Renaissance of the nineteenth century does not readily take to drudgery; it prefers its own conceits to such self-abnegation, and finds it an easier and more remunerative business to play to that insatiable craving for the picturesque which can only end by degrading the profoundest and most permanent of the arts into a mere affair of fashions. It is significant," he continues, "that in his working drawings it was Inigo Jones's custom to make a sketch, and then specify all the proportions of the design in writing at the side. Our habit is to arrive at our proportions in the process of making the drawing."

VI

WREN

VI

WREN

Wren and Oliver Cromwell—Wren and Webb—The Chapel of Pembroke College, Cambridge—The Sheldonian Theatre—The Library, Trinity College, Cambridge—Wren at Paris—The Great Fire—Windsor—Chelsea—Greenwich—The Monument—Hampton Court—Kensington.

WREN was twenty when Inigo died. We have a somewhat pleasing picture of him about this time; but it does not connect him with architecture. He had already, at Cambridge, made himself a name as a mathematical student, had devised an astronomical instrument, and dabbled successfully in poetry. In 1644, Evelyn met him, and describes him as "that miracle of a youth." Miss Phillimore, in her book on *Sir Christopher Wren*, conjectures that he must have met Inigo Jones, whose portico at St. Paul's he spoke of as "an exquisite piece in itself." When he came to live in London, he made acquaintance with Cromwell's son-in-law, Claypole, and through him with Cromwell himself, who offered to release Wren's uncle, Matthew Wren, the old Bishop of Ely, who had long lain in the Tower. But the Bishop would not accept the Protector's terms, and continued a prisoner during the brief remainder of Cromwell's life.

Architecture was at a stand-still; yet it would seem that during the Commonwealth Wren studied it, and with the same earnestness which he brought to every pursuit in which he engaged. Astronomy was, however, his chief object; and he was appointed Savilian professor at Oxford the year after the Restoration, and employed himself in making a lunar telescope. The same year, 1661, he was created both D.C.L. at Oxford and LL.D. at Cambridge. That he was known to have made some progress in architecture is proved by an event which connects his name in an interesting way with that of Inigo Jones. While Inigo was yet alive, the King had given, or sold, the reversion of his office as Surveyor-General of Works, to Sir John Denham. The office was a barren one during the Commonwealth; but at the Restoration many things had to be done for which Denham was wholly unfit. Evelyn characterises him as "better poet than architect." Webb, Jones's pupil, assisted him, informally, we may suppose, and had a promise of the reversion of the office. Webb seems to have died before Denham; and when Charles decided to complete the building of the palace at Greenwich, it is probable that Evelyn recommended Wren to the King. He did not come to the office ignorant of architecture, and had now a great opportunity of putting his skill to the test.

How he had learned to design, we may judge by a piece of architectural work, the first in which he was ever engaged. This was the chapel of Pembroke College at Cambridge. Bishop Matthew Wren, his uncle, on emerging from the Tower, where he had been imprisoned for eighteen years, determined, as a thank-offering, to build a new chapel for his old college, and employed his nephew to make the plans. A sum of £5000 was set apart; but the chapel only cost £3658. The old chapel, repaired, became the library: and the new one was consecrated by Bishop Wren, and appropriately dedicated to St. Matthew on the festival of that saint (21st September), 1665. It was a very interesting building, and of great importance in the history of architecture; for Christopher Wren, following, as well as he could, whatever he may have imbibed from Webb and others of the tradition of Inigo Jones, endeavoured to make it beautiful by proportion alone, and without ornament. In this effort he succeeded completely: the chapel of Pembroke for two hundred years remained a standing protest against the work of architects who, knowing nothing about proportion, sought to cover their ignorance with excessive and unmeaning ornament. Naturally, it was exceedingly obnoxious to the architects of the Gothic revival, who of course professed to see no merit in it whatever. In 1881, the authorities of the College, who had already destroyed their ancient hall, the oldest building of the University, allowed a modern architect to enlarge it. This he did by adding 20 feet to its length, and by stripping the plaster from the exterior. The result is extremely curious, and well worthy of study. What was, from its perfect proportions, one of the most satisfactory buildings in Cambridge, at once declined into an unmeaning rectangular structure, too long for its width and height, but otherwise rather insignificant. The College authorities escaped a greater danger, as one architect had wanted to pull it down altogether. At the reconsecration of the chapel, the service was used which, with certain peculiarities, Bishop Wren had used at the original ceremony.

Meanwhile Wren, in the ordinary course of the business of his office under Denham, had an excellent opportunity of further studying the work of Inigo Jones. He was called upon to repair the ravages of the Puritans at St. Paul's. Already he projected "a dome or rotunda, and upon the cupola, for outward ornament, a lantern with a spring top to rise proportionately." It would be interesting to know what he meant here by "proportionately"; but had his views been carried out, the proportions would probably have been altered by some Gothic architect of our own day, unconscious that proportion has anything to do with architecture. The Great Plague stopped all work in the city; and before it could be resumed, the Great Fire came and ruined the old church. In 1664, Dean Barwick had "laid his own relics in those of his church," as his epitaph said, and was succeeded by Sancroft.

Sheldon, who was at this time Archbishop of Canterbury, was anxious to mark his old connection with Oxford, where he had been Warden of All Souls. He accordingly commissioned Wren to design a University Theatre; for hitherto the "comitia" and the "encœnia" had been held in St. Mary's Church, and the jesting usual on such occasions had been carried on within the sacred walls. Evelyn visited Oxford when, as he says in his *Diary*, the foundations had been newly laid and the whole designed by "that incomparable genius, my worthy friend, Dr. Christopher Wren, who showed me the model, not disdaining my advice in some particulars." It was on this occasion that he found Robert Boyle, Wren, and Wallis in the tower of the schools, "with an inverted tube, or telescope, observing the discus of the sun, for the passing of Mercury that day before it; but the latitude was so great that nothing appeared." It will be perceived that Wren did not, so far, let architecture interfere with the pursuit of his astronomical studies.

His great object in the Theatre, which cost the Archbishop

£15,000, was to cover an area of 70 feet by 80 with a roof, unsupported by any arch or column in the interior. The roof, still one of the most extensive known, was repaired in 1802. The University press was above it in a kind of loft from 1669 to 1713; and books printed at Oxford bore a view of the Theatre on the title-page down to 1759. These views show a kind of louvre or cupola, similar to what Wren had placed on his chapel at Cambridge; and there were large dormers. The turret disappeared later, but was restored or replaced by Blore about 1847. The building is said to resemble in plan the Theatre of Marcellus at Rome. The front, the first in which Wren ventured on distinct architectural features, consists of two storeys of arches and engaged Corinthian columns over a low basement. The sides and back have a rusticated lower storey; and there is a polygonal apse. It was opened in July 1669.

In 1665, the year of the Great Plague, Wren made an effort to improve his architectural education by a visit to France, where great works were at the time in progress for Louis XIV. Previously, his friend Dr. Bathurst consulted him as to some alterations at Trinity College, Oxford. A letter on the new buildings, in Elmes's *Life*, is remarkable for the modesty with which Wren expresses his opinions. In it he anticipates making the acquaintance of Bernini and Mansard within a fortnight. By a curious coincidence, he was also at work at Trinity College, Cambridge, where he built the library which forms the west side of Nevile's Court, of which Dyer says in his *History*. "Here it was that our great master of Palladian architecture, Sir Christopher Wren, surveyed his own work, and was satisfied." This library is over a grand colonnade of Wren's favourite Roman Doric. Wright and

Jones (i. 61) justly say that it is one of the noblest rooms in Europe. It is reached by a staircase of black marble, wainscoted with cedar. "In length, this room measures no less than 190 feet, by a breadth of 40 feet; the elevation being estimated at 38 feet. At the southern extremity, it is terminated by folding doors, which open to a balcony from which we have a pleasant view of the College walks and the river. The floor is paved with square slabs of black and white marble, placed diagonally; the doorways at the two ends of the room, and the fronts of the numerous bookcases on each side, are adorned with a profusion of the most exquisite carvings, in lime wood, which are some of the choicest specimens of the works of the celebrated Gibbons." In a letter, sending some drawings to the College authorities, Wren shows his attention to detail. Of the ceiling he says: "The cornices divide the ceiling into three rows of large square panels, answering the pilasters which will prove the best fret, because in a long room it gives the most agreeable perspective."

This fine building cost about £20,000, and was a long time in hand, not being absolutely completed till the end of the century. Wren gave his services gratuitously as his subscription towards the expenses. The attached columns of the exterior are of the Ionic order. The pilasters of the interior are Corinthian.

Long before the Library of Trinity College was completed, or even seriously taken in hand, Wren had paid his long-promised visit to Paris. It is a remarkable fact that he never went to Italy, and the only dome he can ever have seen before he built that of St. Paul's, must have been that placed over the church of the Sorbonne by Le Mercier. It was commenced in 1629, the architect, who was dead before Wren's visit to Paris, having studied in Italy. It reminds us a little of the dome of the great church at Florence, being rather octagonal than round. The first fine dome in Paris was, no doubt, that of the Invalides, by Jules Hardouin Mansard; but it was not built till after 1680. That of the church of St. Genevieve was only raised in 1764, long after Wren was dead. At his visit, he seems to have been immensely taken by a design of Bernini for the Louvre. He says of it: "Bernini's design of the Louvre I would have given my skin for; but the old reserved Italian gave me but a few minutes' view; it was five little designs on paper, for which he hath received as many thousand pistoles." Bernini's design was, however, rejected by Louis XIV., for one by a comparative amateur, the physician Perrault; and Bernini went back to Rome, where he had already made the long semicircular colonnades of St. Peter's.

Wren in his correspondence says nothing of a dome, nor does he seem to have seen Perrault's design. A curious reflection here forces itself upon our minds. Did Mansard show him any sketch of his future dome of the Invalides, and did Perrault show him the coupled columns with which he was about to ornament the east front of the Louvre? He only mentions Bernini; and Fergusson characterises as vulgar and inartistic the design which Wren appreciates so highly. Of other architects, he gives a few names, but says nothing about their works; and Perrault is not among them.

It is in the letter, published in the *Parentalia*, in which he so praises Bernini, that he makes use of an expression about architecture which has become proverbial, and has been already quoted in this book: "Little trinkets are in great vogue; but building ought certainly to have the attribute of eternal, and therefore the only thing incapable of new fashions."

But as to the dome and the coupled columns, we know nothing. He used the dome, and used it better at St. Paul's; and there too he used the coupled Corinthian columns of Perrault. But of all his work with coupled columns, the finest, unquestionably, is at Greenwich, where he used his favourite Roman Doric with such marvellous effect. It is impossible to believe, therefore, that he learned much, if anything, from Mansard or Perrault; and St. Paul's, as well as Greenwich, must have proceeded from a mind uninfluenced by what he had seen in Paris, but working on lines parallel with those of the great foreign architects.

It will not do, however, to assume that Wren learned nothing by his visit to the Continent. Of course he would have learned more had he been able, like Inigo Jones, to go on to Italy. But, besides architects, he mentions Van Ostal and Arnoldin as "plasterers who perform the admirable works at the Louvre"; and Perrot, who is famous for basso-relievos: nor does he overlook the tapestry works in the Rue Gobelins, and "Mons. de la Quintinge," who "has most excellent skill in agriculture, planting, and gardening." In fact, he proposed, when he returned home, to perfect some observations on the state of architecture, arts, and manufactures in France: a work he, probably owing to the progress of affairs at home, never had time to accomplish.

For very soon after he returned, an event occurred which was to send the whole current of his thoughts and energies into one direction for the rest of a very long life; namely, the Great Fire of London. He had come back early in the spring, and was busy on the reparation of St. Paul's. He had been appointed member of a Royal Commission for the purpose, together with John Evelyn, the Bishop of London, and the Dean of St. Paul's. The steeple was pronounced to be in a dangerous condition; and Wren, strongly supported by Evelyn, proposed to rebuild it upon new foundations with a noble cupola: "a form of building previously unknown in England." But before anything more could be done, at ten o'clock one hot Sunday night in September the Great Fire broke out. The next night it took hold of St. Paul's, greatly helped by the wooden scaffolds. On the 7th, five days after the outbreak of the fire, Evelyn penetrated to the City, and found Inigo's beautiful portico rent in pieces, and almost the whole church, except the extreme east end, injured more or less seriously.

It may be worth while here once more to combat a silly idea which has gained currency during the prevalence of the Gothic revival; namely, that Inigo Jones in his portico, and Wren in his projected dome, were doing anything to spoil old St. Paul's. Of Inigo's portico I have said perhaps enough. It was certainly not in any way incongruous, as shown us in Hollar's well-known print in Dugdale's *St. Paul's*. The whole nave, especially as he recased it, belonged more or less closely to the Romanesque style. In the case of the proposed dome, we do not know so much. Wren might either have made it to agree with the east or the west end, with the Pointed or the Romanesque half of the church. It is, in fact, a pity that none of his employments led him to expand, on a larger and more important scale, his beautiful Tom Tower at Oxford. A dome to old St. Paul's in that style would have been in many respects the most beautiful building in England; and, whether contrasted with the somewhat stiff

Perpendicular of the east end, or with the Corinthian portico at the west end, would have added to the beauty and interest of every view of the Cathedral.

There is no occasion to dwell on the terrors or destruction of the Great Fire. We are only concerned here with the fact that it gave Wren the opportunity of displaying the resources of his genius. After a long and careful consideration of his work, after comparing it with that of his predecessors and contemporaries, after throwing into the comparison the best designs made since his time—I have come to a very simple conclusion. Whatever Wren did he did as thoroughly as possible. His genius, it has been said a hundred times, consisted in taking pains. He thought out each problem as it was presented to him. His minor designs have their proportions as carefully fixed as those of the more important. Nothing is neglected that will enhance the effect. Constantly obliged to study cheapness, he made up for it by spending thought. His slightest design was mixed, to apply John Opie's phrase, "with brains." By the constant use of this ingredient, cheapness was ennobled.

Among the poorest buildings in materials and size, we find such beautiful "bits" as the east end of St. Peter's upon Cornhill, or the exquisite but simple little tower of All Hallows, Bread Street, lately pulled down. The destruction of many of his City churches—a subject to which I shall have to revert in the next chapter—has shown how careful he was that each tower should have its place in relation to the towers nearest to it; and what irremediable damage has been done, by the removal of even the smallest and poorest of them, to the general effect of the whole number.

GREENWICH HOSPITAL.

We had better examine Wren's domestic and public buildings first, and his City churches in a chapter by themselves. His library at Cambridge has already been described; but in addition he built palaces, hospitals, town halls, and private houses, all characterised by the same qualities of design and execution. The maximum of beauty, the minimum of cost, combined with the utmost stability—those were the objects at which he chiefly aimed, and aimed with success.

It is not possible to range Wren's buildings of this kind in chronological order; because most of them were undertakings of such magnitude that they went on, as I might say, perennially during a long life. Such were Greenwich Hospital, for which he made designs in or before 1695, but which was not completed till after his death; or Hampton Court Palace, where he first went to work in 1689. He was also employed at Windsor Castle, and built the Town Hall of the royal borough. He designed, in the plainest style, the hospital for soldiers at Chelsea, the building of which went on from 1682 to 1691. But the finest of all his buildings of this kind is neither a palace nor a college,—it is the seamen's hospital at Greenwich. No one who has an eye for stately scenic effect can help feeling an enthusiasm for this beautiful work of art: the finest of the kind in Europe. Marlborough House was designed in 1709. It has had storeys added, which make its proportions no longer what Wren designed. Kensington Palace, as we now see it, is also mainly his; monograms and other devices of his period being on the east and north-east side: and the Orangery, a beautiful but neglected building, as well as an alcove, now removed, were designed by him for Queen Anne. The Monument, Fish Street Hill, simple as it looks, cost Wren a great deal of thought and care. He began it in 1671.

He built smaller palaces for Charles II. at Newmarket and at Winchester. The latter survives as a barrack, and though devoid of ornament, is of excellent proportions, which are enhanced by contrast with those of some later buildings adjacent. It is curious how few private houses seem to be his; but it is impossible, either at Salisbury or Chichester, not to see his hand in some of the residences connected with the Cathedral. At Salisbury, there is a beautiful stone-fronted house on the north side of the Close, which must be his. It bears the initials C. M. and the date 1701. Another house, on the west side, also looks like his work. It is raised on a high basement, and consists only of one principal storey, in red brick with stone corners. At St. Albans, an almshouse near St. Peter's Church is probably his; the probability being increased by the fact that Edward Strong, his master mason, is buried in the church. It is the epitaph on Strong's monument that gave rise to the saying that St. Paul's was built by one architect, under one bishop, and by one master mason. The inscription, however, only says that Strong, in company with Wren and Compton, "shared the felicity of seeing both the beginning and finishing of that stupendous fabric." This is likely enough, as his elder brother, Thomas Strong, was the first master mason; and Compton, who was then Bishop of Oxford, was much about the court, where he was tutor to the two princesses, afterwards Queens Mary II. and Anne.

It would be easy to make a large volume about Wren's domestic architecture; but it will suffice here to notice a few only of those examples which may be singled out as having advanced taste, and added to the charms we are but now beginning to recognise in the "Queen Anne Style." If proportion was the ruling motive of Inigo Jones, it was still more so that of Wren.

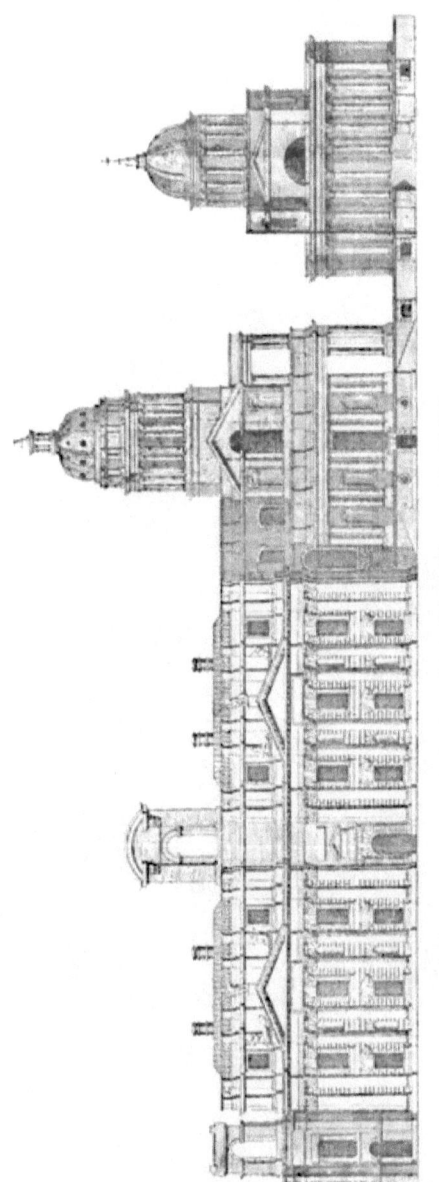

PART OF WREN'S FIRST DESIGN FOR GREENWICH.

Many of his buildings were absolutely devoid of ornament, and this, too, even at the beginning of his career. Chelsea Hospital was begun in 1682. Mr. Beaver, the latest historian of Chelsea (p. 277), thus speaks of it: "With very simple materials, Wren has contrived to give us a building perfect in proportion and dignified in effect—one on which the eye dwells with pleasure. Its regularity suggests no monotony; its simplicity, no poverty of design: it bears the stamp of genius." Another writer, Mr. L'Estrange, in his *Village of Palaces* (ii. 22), says of it: "Few can pass before this noble pile of buildings, without being impressed with the size, strength, and symmetry of Wren's design." For it Wren is reported to have received £1000: for the design of Greenwich Hospital, he refused remuneration, making his work a contribution to the charity, just as at Cambridge he had made his old college a present of his design for the noble library.

Greenwich must at first have presented a very interesting problem to his mind. The conditions were as follows: The old palace of the Tudors has been pulled down; a few fragments, chiefly of outbuildings, only surviving. A new palace, in a very stately style, by Webb, Inigo Jones's pupil, had not very long been built, and was still in parts incomplete, though Charles II. is said to have stayed for a time in the eastern wing. The plan was oblong; a narrow court in the middle being entered by a low archway with a wider opening to the south. The western side was, if complete at all, only in brick, and was finished in stone by George III. The front towards the river had two porticoes, formed of four engaged Corinthian columns, supported at the corners by pilasters of the order, with a noble projecting cornice, very much in Jones's style.

Above this is now an attic storey, which, I presume, was an addition perhaps by Wren. On the eastern side the palace had a single central portico, and had the corners accentuated by pilasters. The gardens, or grounds, on this side were cut close by a broad walk or road which led from the bank of the river to a house some way off, nearer Greenwich Park. This house had originally been built for Henrietta Maria, by Inigo Jones. The basement remains, and possibly the semicircular stairway to the terrace may be of his time: otherwise, there are no architectural features left.

The problem before Wren was how to work in the palace of Charles II.; to make use of as much of the vacant space as possible between it and the Queen's House; to arrange so as not to stop the road from the river; to provide for a possibility of increase of accommodation equal to at least three times that of the existing palace; and to arrange the new buildings with the old in such a way that they should not, so to speak, hide each other.

All these objects he accomplished. A second building, on the east side of the road, was erected with one front to the river and another to westward, looking on the older palace, in exact, or almost exact, imitation of what Webb had built for Charles. This new "pavilion" was called after Queen Anne, in whose reign it was completed. As it was built on a line with the palace, a place where there was a kind of bay of the river, beside which stood the old chapel of the Tudor palace, was filled up, and the new foundations put upon it. Behind these two buildings, those namely of King Charles and Queen Anne, were placed two other rectangular blocks, in the western-most of which was the great dining-hall, and in the easternmost the chapel. These blocks were brought forward, almost to the Queen's roadway, and were edged on the sides facing each other with a magnificent colonnade of coupled pillars of Wren's favourite Tuscan Doric. This colonnade shows six pairs of its double columns on either side, and a long vista of 150 pairs towards the gates of the Queen's House and the Park. The effect is magical, and is greatly increased in symmetry and beauty by the domes which stand on either hand above the entrances of the chapel and the hall. If Wren had never built another dome, this pair, though not large, should have immortalised him. Coupled Composite columns are employed; and there are " four projecting groups of columns at the quoins. The attic above is a circle without breaks, covered with the dome and terminated with the turret." This is from the official account published in 1789; but no words can give an adequate idea of the effect.

Unfortunately, the works were never finished by Wren. On the west side of King William's building, a very different hand was employed. Vanbrugh is responsible for the heavy brick with coarsely-moulded stone dressings which is such an eyesore in one of the most prominent views of the Hospital. In the reign of George III. some similar work in the building of King Charles was cased in stone, but in a poor and heavy style; and altogether, this western front, the first to greet the traveller from London by land, is the least beautiful part of the whole composition. Vanbrugh's work, of which there is more in the neighbourhood, should be a warning to those modern architects who seem so anxious to imitate his anomalous style. The dates of the different parts of Wren's work are worth noting. The hall and chapel were founded in 1696, and opened in 1705. Queen Anne's building was founded in 1698, but was not finished till 1728.

Not far from Greenwich is another very dignified front by Wren: Morden College, Charlton. It may be compared with the almshouse at St. Albans; but is on a larger scale, and there are more architectural features in the courtyard with its "piazzas." Another small work in this grand but simple style is the gateway of the Middle Temple: a building little noticed till lately, but now forced into prominence by the contrast it presents, in its simplicity, proportion, and good taste, to the newer buildings which surround it. Temple Bar was a beautiful stone archway in Wren's best manner. It was pulled down in 1878 for reasons which have not transpired. Some said it obstructed the traffic, which cannot have been the case. It would have been easy to take the roadway past it on one side, as has been done with old gateways in Paris. The north side was open at the time. That it obstructed the traffic cannot, however, have been the reason for its removal; as a monument, hideous in design and nearly as obstructive, has replaced it in the middle of the street. Temple Bar has been made the entrance gate of Theobalds Park in Hertfordshire, and looks extremely well there, surrounded by old oak trees; but its loss as a dignified, beautiful, and historically interesting entrance to the City is irreparable.

Wren spent a great deal of thought on the Monument. He first intended it to be an unfluted Tuscan Doric column, with flames of gilt bronze issuing from it at intervals, and on the top an urn, flaming, with a phœnix rising from it. "On second thoughts," he says, "I rejected it; because it will be costly, not easily understood at that height, and worse understood at a distance, and lastly, dangerous, by reason of the sail the spread wings will carry in the wind." A good engraving of this design was made by Hulsbergh, who also issued a large print, from a drawing by Hawksmoor, of the final design. This represents a fluted column of the same order, with a large "egg and dart" moulding to crown the capital, and a statue of Charles II. on the summit. An urn was substituted for the statue, "contra architecti intentionem," says Hawksmoor. There are enlarged views of the plinth and other details in the print, which show with what care and attention each part was considered and thought out. The pillar was completed in 1677.

Wren began to work at Hampton Court in 1689, and continued to superintend the alterations and improvements until 1718, when, by means of an unworthy intrigue, he was dismissed from his office of surveyor. A very interesting design of his is printed by Mr. Ernest Law in the third volume of his *History of Hampton Court*. The original is still in the Office of Works. It seems to have been made in 1699. The avenue of horse-chestnuts in Bushey Park was to be made a grand approach to the palace on the north. Two wings, 300 feet long, were to be built from the great hall, with colonnades; and the intervening offices were to be cleared away. The result would have been to make another grand scenic effect like that at Greenwich, and to give William III. a palace superior to any other in Europe. But the King died; and though Queen Anne was much at Hampton Court, nothing was done to complete what Wren had begun. As we see it now, the mixture of Gothic and Palladian is so charming that we cannot wish it otherwise; but, no doubt, from an architectural point of view, its incompleteness is to be regretted. The east front, as finished by Wren, is very fine, as is the Fountain Court; while the Ionic colonnade in the Clock Court adds greatly, by contrast, to the picturesque effect of this part of the palace. The carvings everywhere are extremely good, and are worthy of separate and careful examination. The architect's initials occur on the west side of the cloisters which surround the Fountain Court; and it is conjectured that they mark the lodgings he occupied while he was engaged here. Colley Cibber seems to have been the principal sculptor employed; but there is occasional mention of "a Frenchman." The east front is of red brick, with stone dressings; and, except for four Corinthian columns and a pediment in the centre, is almost devoid of ornament.

Wren built much in a very plain style for William III. at Kensington. Work was begun as early as 1689, but was interrupted by a fire in 1691. In 1696, Evelyn records a visit to it, and specially mentions a gallery and "a pretty private library." A good deal was added to the palace by Kent for George II., so that, as also at Hampton Court, it is not always easy to distinguish Wren's work; but we can be sure of such parts as bear the initials of William and Mary, or of Queen Anne. As originally laid out, there was a formal garden south and east of the house. North of it was a lawn on which some cedars grew; and apparently there was a bronze bust or statue among them. On this side also was a long walk or avenue leading to the Bayswater Road. Its southern end is marked, near the Orangery, by two fine brick piers, with carved stone ornaments. Here, it is said, William used to walk alone when in a moody humour. Looking southward is the Orangery: and the principal walks of the flower garden converged on the Alcove: a beautiful feature, which stood facing the Orangery at some distance, its back being against the brick wall, which at that time shut off the Kensington high-road. In this Alcove, it is said that the French refugees celebrated a mass in fine summer weather. When the wall was taken down, the Alcove was senselessly removed, and placed at the other end of the gardens. It was designed to face northward, and to form the end of a vista. It is now placed facing south, about half-way up a slope, where there is no vista, and where it only serves to make more hideous the buildings about the head of the Serpentine close by. The Alcove bears the initials of Queen Anne.

The Orangery is left in a melancholy state of neglect. Here Wren did his best to decorate a garden with an ornamental building; and we see from it what he could accomplish in a style more playful than that he usually employed. The columns of the central bay are of red brick, of his favourite Doric order; and there is very little ornament anywhere. The result, though simple, is eminently satisfactory. It is sad to see this beautiful building used as a kind of tool-shed for the gardeners, a place for mixing manures, with a series of squalid hothouses obscuring the best view of its front.

VII

WREN'S CHURCHES

VII

WREN'S CHURCHES

Obnoxious to bishops — Many destroyed — Method of procedure — Case of St. Antholin's — A monstrous falsehood — Classification — St. Paul's — Court influence — A Protestant design — An artificial design — Decorations — Parish churches — Two principal patterns — Domed churches — Gothic churches.

FUTURE historians of the architectural movements of the nineteenth century will be sufficiently puzzled to account for the rise of the anomalous or eclectic style to occupy all their faculties. But if any time remains, they may inquire into another and still more surprising phenomenon. It is well known among foreign nations, though apparently not among ourselves, that in Wren's City churches England possesses, or, to be accurate, possessed, a treasure only comparable to the works of art at Florence and Rome: a treasure such as no other city could show. Yet, incredible as it may seem, those in authority have for more than thirty years past been making the most strenuous efforts to destroy these treasures. It is difficult, or rather, impossible, to find the reason for this course of action. Some years ago, I endeavoured to account for it by the action of superstition; but I am assured now that I was mistaken. I can, of course, understand that the architects who are transforming the City would be glad to remove such prominent witnesses of their own incompetence. But the churches have not been pulled down at the instance of architects: they have been removed at the instance of successive bishops and other ecclesiastical persons. This seems to me the most extraordinary part of the story, though, considering the character and position of these personages, there is another fact almost as extraordinary. The churches have been condemned with the consent of the parishioners; and this consent has been obtained by means of deception. I do not mean to impute this deception to the ecclesiastical dignitaries just mentioned; but they have profited by it. The method of procedure has been briefly this. When a church was to be destroyed, the parishioners were informed that it was not designed by Wren at all; or, failing that assertion, they were told that the church Wren had designed for that parish was pulled down long ago, and the present church built by somebody else in imitation of it. This course was repeatedly pursued. The church was subjected to Jedburgh law. It was condemned and pulled down first, and judged and acquitted afterwards. The whole story was told lately of one of the most precious of these churches in a letter in the *Times* (21st April 1892), written by the churchwarden who had been made the cat's-paw of the religious functionaries I have mentioned. He now bitterly laments the fraudulent part he was deceived into playing. People have so little architectural taste, and so few of the people of a parish in the City knew whether their church had an artistic value or not, that these tactics have been marvellously successful.

It is therefore necessary once more, and as many times more as possible, to reiterate the fact that all Wren's churches in the City were designed with a purpose, and that the destruction of one church is a partial destruction of all the rest. This is a consideration which cannot be too strongly insisted upon. As a rule, the relation between the different churches was preserved by the spires or towers. If a church is removed, it may be sufficient to leave the tower. An inquiry ought to have been made in every case to this effect; but, so far as the general public is informed, no such question has ever been raised. Church and church tower have both been destroyed in nearly all cases; but an agitation, got up in time, was successful in saving the tower of St. Mary Somerset, the removal of which would have had an even more disastrous effect on the view of London from the Thames than the destruction of St. Antholin. It must be remembered that the decree of condemnation is still in force, and as I write, for aught I know to the contrary, may be in process of execution. The right reverend and reverend society for the suppression of Wren's churches goes to work with exceeding subtlety; and it is certain, that if it put as much skill and craft into motion to obtain the money by other and less nefarious means, the task would be comparatively easy, and the members would be saved the necessity of absolving their agents from the sin of mendacity. I do not believe one more of Wren's churches would be destroyed if the Bishop of London could be made to understand that the consent of the parishioners can only be obtained by simple lying.

One fact is worth a great many arguments. I will offer the reader two facts. Wren built a very curious and interesting church, called St. Mary Aldermary, in a modification of the Gothic style. The chief feature of this church is the beautifully proportioned tower, in which some people see an imitation of the tower of Magdalen College at Oxford. By way of contrast,

Wren built over against St. Mary's another tower: that of St. Antholin, Watling Street. The tower of St. Mary's is 135 feet high, having corner pinnacles. The spire was designed to be a little taller, as suitable to its form; and it rose to 154 feet. It was very much like a Gothic spire, and was built of stone; in this respect differing from all but one other of Wren's spires. But though so Gothic in its general form, it was strictly Palladian in details. The harmony, or contrast of the two steeples—for they stood very close together, and you could hardly look at one without seeing the other—produced on the mind of any one of artistic taste a feeling of intense pleasure: a distinct thrill, like that produced by beautiful music. When the new street was made, these two towers stood on either side of it, and opposite to each other. In those days, I was obliged to spend six months of every year abroad; and I well remember making an exertion always on my return to go and have a look at the pair, sometimes endeavouring to group St. Mary-le-Bow or some other tower with them in one view. Judge my distress, in 1877, on returning from a winter in Egypt, to find St. Antholin's gone, and its place occupied by some shops, rather conspicuous for the poverty of their architectural features. The miserable story leaked out by degrees. When the ecclesiastical authorities first proposed to destroy St. Antholin's, a cry of horror and indignation went up. But they were not to be balked of their prey. They discovered, or allowed some one to discover for them, that St. Antholin's was not designed by Wren,—that, in fact, it had only been built a few years,—and that it was by no means worth the fuss being made about it. I do not say that the authorities believed this tale themselves: they are gentlemen of education and must have known better. But they did not on

GREENWICH: VANBRUGH'S WORK.

that account contradict the story. By much canvassing and many reiterations, a bare majority of the parishioners was obtained. Even then, a number pleaded for the reprieve of the spire ; and it was spared for a few months, but, to use the words of Mr. Andrew Taylor (*The Towers and Steeples of Sir Christopher Wren*, p. 38), "the increased price which was thereby obtainable for the site finally outweighing all less mercenary considerations, it shared the fate of the church ; and the place that once knew it knows it no more for ever." I was lately assured by an alderman, at that time Lord Mayor, that the whole bench of aldermen protested in vain against the removal of this tower. Then came the sequel to the story. The gentleman who had been made a cat's-paw found out that he was wholly deceived. A few courses of the spire had on one occasion been taken down in order to remove a faulty piece of stone, and had been scrupulously replaced. It was upon this repair that the whole monstrous lie told to the parishioners rested. I repeat, I cannot acquit the members of the Episcopal committee of blame.

During the present year, nevertheless, the same tactics were tried in order to destroy Wren's only other stone spire, for of course the steeples of St. Mary-le-Bow or St. Bride's are not exactly spires,—but the parishioners of St. Antholin, still smarting under the misfortune of 1876, told the whole story as I have endeavoured to narrate it above, with the result that for the time being St. Dunstan in the East is saved. The admiration of the citizens, and of all people of any taste, had not been able to save St. Antholin.

It had been built in 1682 ; and the cost of church and spire was £5700. St. Mary Aldermary was built at the same time, and finished only a few months sooner ; so that there can be no doubt of Wren's intention of making one composition of the two.

It is not necessary to go through all Wren's City churches. The following have been pulled down by the bishop and his assessors within the past few years: St. Antholin; All Hallows, Thames Street; All Hallows, Bread Street; St. Mildred; St. Michael, Queenhithe; St. Dionis; St. Benet, Gracechurch Street; and St. Olave, Jewry. Besides these, St. Christopher by the Bank; St. Michael, Crooked Lane; and St. Benet Fink, were previously pulled down, but are not so much to be lamented. In all, ten of Wren's churches have been destroyed under the "Union of Benefices Act" of 1860; and Mr. Taylor says that "under a scheme drawn up by the Fellows of Sion College in 1876, thirty-one more City churches were marked for destruction. This, however, was too much even for the apathy of the British public; and steps were taken to resist such a scheme, which were so far successful that the matter has been allowed to drop for the present; but it may be resuscitated at any time; and it is imperatively necessary, therefore, that a greater public interest be awakened in the churches, that we be not implicated in deeds for which posterity will not hold us blameless."

Various attempts have been made to classify Wren's churches. They generally end in leaving the classification like that of Greek verbs. One is regular and all the rest are exceptional. We cannot class St. Paul's with any of the others, though the interior of St. Stephen, Wallbrook, may be compared with parts of it. There are two or three fine churches which go together: St. Lawrence Jewry; St. James, Piccadilly; St. Bride, Fleet Street; and St. Mary-le-Bow—in all of which Wren showed his marvellous skill in covering a wide space where all can see and

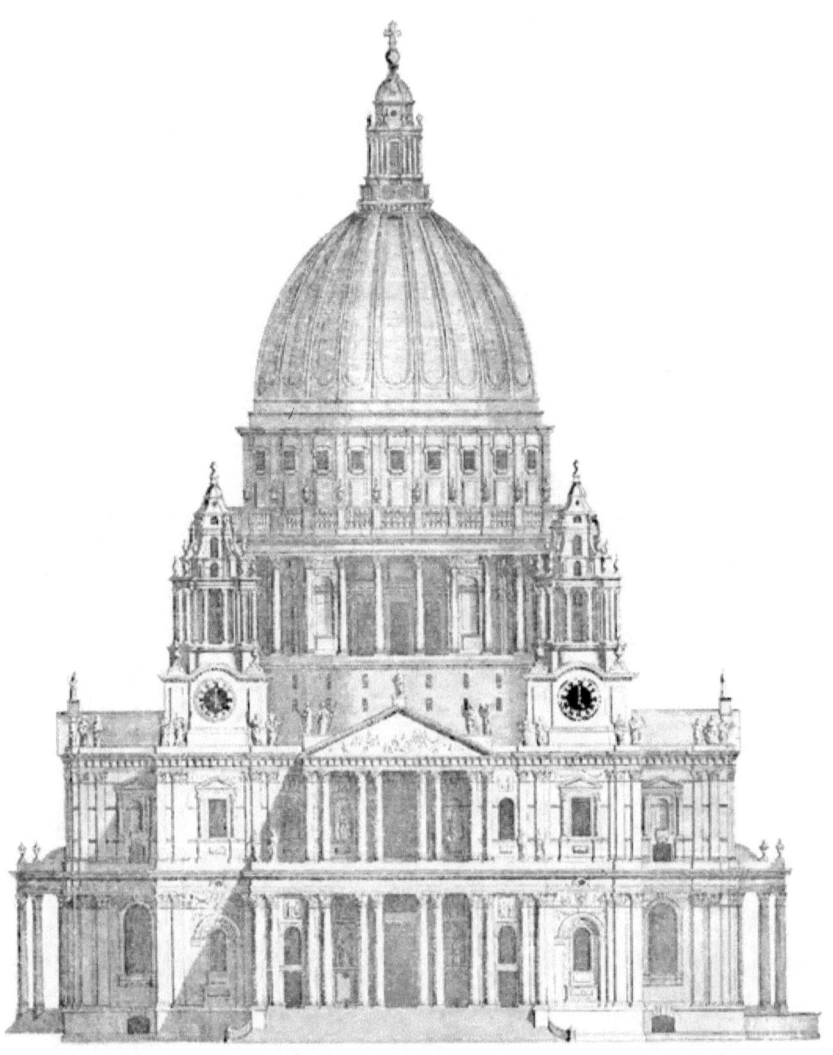

The West Prospect of ST. PAUL'S CHURCH, begun Anno 1672 and finished 1710. From an old print.

hear. Again, in some of the smaller churches, where cheapness had to be considered before all else, he contrived to bestow some feature, within or without, which carried off the plainness he could not otherwise avoid. Mr. Taylor classifies them by the towers and spires alone. Some have stone steeples, like St. Mary-le-Bow, Christ Church, St. Bride, and St. Vedast. Some have lead spires and lanterns, of which type St. Magnus and St. Lawrence are good examples. A third kind have square towers, like St. Andrew by the Wardrobe, or St. George, Botolph Lane. Finally, we have the Gothic churches or church towers: a very interesting class, which comprises St. Mary Aldermary; St. Dunstan in the East; St. Michael, Cornhill; and St. Alban, Wood Street.

Of St. Paul's, a very short notice must suffice, and that chiefly by way rather of praise than of criticism. But it is well to point out the conditions under which St. Paul's was designed. Let us ask ourselves what were the objects of church-builders in the second half of the seventeenth century. The question is wholly one of church doctrine. The school of Laud had passed away, and its leader had been one of the victims of the Great Rebellion. Of his school, we have a most interesting example in St. Katharine Cree, a church probably built by Inigo Jones, under Laud's personal superintendence; that is, it was designed by a Roman Catholic for a bishop whose leading idea was union with Rome. I am expressing no opinion on the religious questions involved, but only touching on them as they affected architecture. In St. Katharine Cree, accordingly, we see a building designed for the celebration of Mass. The designer's object was, briefly, to construct a church in which every worshipper could see the elevation of the Host. But when the Revolution, which had been fostered and promoted as much by the tendency of Laud just mentioned as by any one thing besides,—when the Revolution had swept over the land, views and opinions on these subjects were wholly changed. During the Commonwealth, in the City of London at least, every parish had chosen to itself a lecturer or preaching clergyman. Under the old rule, before the Reformation, and long afterwards, preaching was no part of the duty of the parochial clergy. They went on celebrating Mass until that was forbidden. Queen Elizabeth licensed a few preachers from time to time; and they chiefly held forth at such a place as St. Paul's Cross, and were strictly amenable to the authorities. In this respect, there was more liberty under James I. and Charles I.; and then, as I have said, the people grew so fond of sermons, that they elected preachers. During the reign of Cromwell, many parish clergymen fled and left their people to the lecturers, who subsequently showed well by contrast during the prevalence of the Great Plague. It is probable that, if we had Mr. Besant's powers, and could go back and interview a citizen of London about the beginning of 1666, and could ask him "What is a church for?" he would reply, "It is a place where we can hear sermons." So when, later in the same year, the Fire came and burnt nearly all the old churches,—churches, we must remember, built for Mass, not for sermons,—the citizens, in rebuilding them, thought only of how they could hear and how they could see, not the elevation of the Host, but the face of the preacher. Wren says himself in *Parentalia* that his object in designing St. James's, Piccadilly, was to make it "so capacious, with pews and galleries, as to hold above 2000 persons, and all to hear distinctly and see the preacher." This, then, was the leading idea, the motive of the church-designer of the latter half of the seventeenth century.

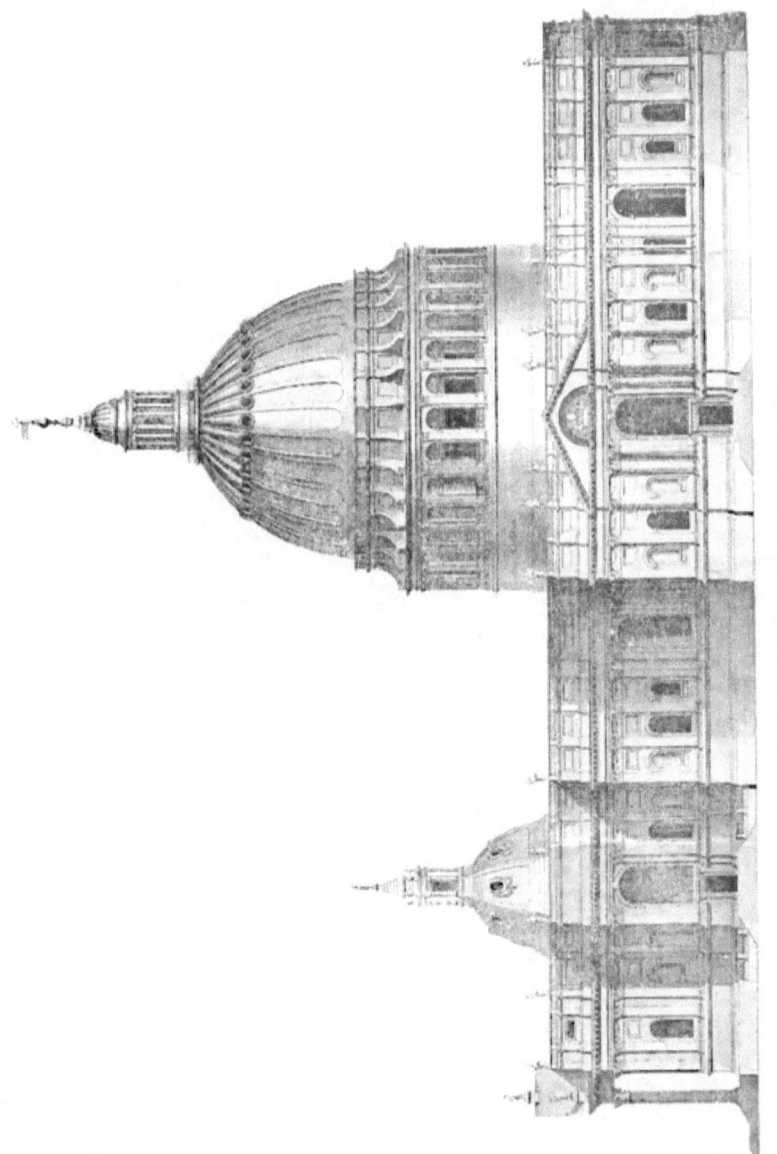

ST. PAUL'S CATHEDRAL: WREN'S FIRST DESIGN

There was, however, another influence at work, and that too in circles where it could not be ignored. The king's brother and next heir, the Duke of York, was a fervent Roman Catholic. He hoped, if ever he succeeded to the throne, that he should be able to lead England back to the true faith, and to see the Mass celebrated everywhere. The ecclesiastical question was thus complicated. James could not impose his views on the parishioners of Piccadilly, or of any part of the City; but St. Paul's was a building of public importance: it was the cathedral of the greatest diocese in England; and it was by the king's liberality, but much more by the exercise of his prerogative, that Wren expected to obtain the money necessary for his gigantic undertaking. This being the case, the voice of the court prevailed easily over the voice of the city. The citizens knew what they wanted; but they were too much impoverished by the Fire to be able to give handsomely to St. Paul's. Most of them, too, had their own churches to think of and to rebuild. There are a few cases in which a citizen gave £1000 to his own parish church and the same to St. Paul's; but whereas £1000 was nearly enough to build a parish church, it was but a mite in the subscription list of the Cathedral. But the citizens knew what they wanted; and Wren was well acquainted with their wishes. They desired a vast preaching-house; and Wren designed them such a glorified preaching-house as the world has never seen. It may be worth while to attempt a short description of it.

The principal feature was a dome. Round this dome were aisles, or an aisle, of great width; the only interruption to the sight being the piers supporting the dome, which were divided and reduced in thickness to the utmost, so as to leave the whole vast space under the dome and aisles as free as possible. Miss Phillimore (*Sir Christopher Wren*, p. 197) does not seem quite to understand the object of the design. She says: "The ground-plan was that of a Greek cross; the choir was circular; it had a very short nave, and no aisles." It is not easy to understand what is meant by a circular choir. At the extreme east end, Wren made a kind of semicircular recess: a "Kibleh," to borrow a word from Saracenic architecture. The eastern and western limbs were of the same length precisely as the northern and southern. We may talk of them as choir and nave, and as transepts; but the architect's one object was evidently to provide the largest possible open space, call its parts what you please, and to make no provision whatever for Roman Catholic worship, for Masses in side chapels, for a high altar, or, in short, for anything but a place for all "to hear the service, and both to hear distinctly and see the preacher." In order to add to the room, and also, no doubt, in order to obviate any monotony that might arise from the grouping of the aisles round a circular centre, he made the exterior walls of his aisles convex. Here Miss Phillimore's remarks are excellent, and I venture to quote them: "The outside, with the two hollow curves joining the transepts with the nave, and the two different-sized domes, would probably have been disappointing: but one speaks with diffidence, for this was Sir Christopher's favourite design—the St. Paul's which he told his son he would most cheerfully have accomplished." It is possible, however, that the hollow curves were intended to let the dome be better seen.

The second dome mentioned here was one of the most curious features of the design, and like the "hollow curves" tries our faith. At the west end of the church, instead of the two noble campanile towers, there was to be a portico very like that of Inigo Jones, destroyed in the Great Fire. There was only one range of columns, not two as at present. Through the portico we were to enter a vestibule. This vestibule was to consist of a narrow, windowless bay, a wider one, circular and domed, and a third like the first, and then came what Miss Phillimore calls the nave; for this vestibule was no part of the church, but was suitable as a meeting-place for the chapter and such purposes. One thing is conspicuous by its absence. There is not a vestige of a chapel: not even such open chapels as those north and south of the entrance to the nave as it now is. In short, Wren's one object was to provide his friends the citizens with what they entirely desired, a grand house in which to hear sermons.

But, as I have hinted, other influences prevailed. Supposing England to become once more Romanist, a cathedral in which not only was there no provision for chantries and chapels, but in which no such provision was possible, would be useless. The design, accordingly, which Wren carried out in a beautiful wooden model, still extant, though much injured by the neglect of successive deans and chapters of St. Paul's, its custodians, was found unsuitable by the court party.

It is said, that on learning the decision of the King and the Duke of York, Wren endeavoured to alter it in vain, becoming at length so agitated as to shed tears. A new design was to be produced. We can well imagine that Wren felt angry. Weale has some appropriate remarks on this conjuncture (*London*, i. 182): "No perplexity that can assail an architect can well equal the difficulty of Wren's task, between a Protestant nation and a Catholic future monarch, to plan a temple that might upon occasion serve for either religion, and therefore for neither well."

Undoubtedly, he took some pains with the new design. It was not in his nature to make any design without more or less pains. Taking the lines of an old cathedral, such as St Paul's had been before the Fire, and reducing the confusion of styles and parts to proportion and harmony, he made a drawing for the King and the Duke. "Many a deep study," says Weale, "had to be wasted, many a beautiful invention abandoned, before he could descend to a design sufficiently tame and commonplace to meet their notions." One of these is probably a design exhibited in an unsigned and undated engraving, entitled, "Section of the Cathedral Church of St. Paul, Lond: wherein the Dome is represented according to a former Design of the Architect, Sr Chr. Wren, K^t." The west end is in two storeys, as at present; two chapels, right and left, are introduced at the extremity of the nave, but there is some awkwardness about the arches which lead into them. There is a long choir terminating in an apse. The great central dome is in its place; but we see no western towers, and no secondary dome. Perhaps it is to this design that another anonymous engraving belongs. It is labelled " Elevation of the West Front of St. Paul's Cathedral, Lond: according to the former Design for the Towers." Any one who sees this drawing for the first time will rub his eyes. It appears to be without the smallest spark of Wren's genius about it. The front, with its rows of coupled pillars, was to be as it is now, but flanked by two pepper-box turrets, with little cupolas rising from a tame colonnade of plain pillars. The one touch of Wren about all this design was to be seen in the position and supports of the dome. Before we go on to speak of it, we may as well look into the rest of the miserable history of the interference of Charles and James, and its ultimate effect on Wren's design.

Before finally giving up hope, he seems to have endeavoured to show that an east end of the most "advanced" character might be constructed in his original design. A drawing is reproduced by Longman in his *Three Cathedrals* (p. 111), who, however, gives no authority for it, from which it would appear that Wren proposed to fence in the easternmost bay so as to produce what probably led Miss Phillimore to speak of a "circular chancel." Longman also reproduced a design for the east end, which Wren proposed to fit on to the model he had made. But nothing satisfied the Court party, into which all the clergy who held high church views were adroitly drawn. Wren made drawing after drawing, until at length, in utter despair, and in the nearest thing to a bad temper of which his meek and quiet spirit was capable, he submitted to the King a design which makes the student who sees it for the first time laugh involuntarily. It is impossible not to think of a Burmese pagoda. To say it is ludicrous is to understate the case. It is impossible. How Wren must have chuckled to himself while Charles wrote on the sketch, "We found it very artificial, proper, and useful." But Wren had no idea of ever carrying it out, or anything like it. He wanted the King's signature to a design, and now he had obtained it. A little experience, not only as to St. Paul's, had taught him that the King had no taste, and that the Duke of York had, if possible, less. Longman describes the drawing as "poor and tawdry." Miss Phillimore gives an amusing account of it, pronouncing it "artificial" in the modern sense of the word. The west end was to be like that of old St. Paul's as Inigo Jones left it. "There is a low flat dome, then a lantern with ribbed vaulting, surmounted by a spire something like St. Bride's, but thin and ungraceful." Elmes, Fergusson,

Weale, and Garbett make no mention of it; and, in truth, it would be absurd to treat it seriously. But Wren had exactly gauged his royal master's inclinations, and by means of this caricature of a design, he pushed himself further into the King's confidence than he had ever been able to do by his best drawings. Charles added to his approval of this "artificial" cathedral, leave to make such alterations, "rather ornamental than essential," as from time to time he should see proper, and furthermore left the whole absolutely to Wren's management. This was in May 1675; and in June the first stone was laid. The design, which Wren reserved, *in petto*, for the time being, was even more unlike his first two-domed drawing than it was unlike the "proper and artificial design." Instead of the tall western portico, the difficulty of finding stones of sufficient size for the large columns of a single range obliged him to adopt two ranges, each of half the size. Two western towers, as they were gradually built, showed the perfection of beauty in form and proportion and contrast with the dome. The dome itself, instead of being like a Burmese pagoda, grew into the noblest dome in Christendom, and the greatest ornament of the greatest city in the world. All these innovations on the design approved by the King were made silently; Wren apparently keeping in his mind as dominant ideas, first, the Gothic plan, and secondly, a particular modification of that plan, which, no doubt, he had observed and studied when he visited his uncle the Bishop, at Ely, where, by the way, his hand may yet be traced in some improvements at the palace.

This is not the place for a detailed account of St. Paul's as we see it. I am glad to think it has survived the so-called Gothic revival, which more than once threatened it with injury and even destruction. The dean and chapter have always had

the command of too much money, and where taste has been
wanting, they have nevertheless, on account of their wealth, been
able to do things of which no one who has studied the subject
can approve. The removal of the choir-screen and organ may
be mentioned. I am not inclined to condemn the reredos,
which is very handsome; but it is sadly wanting in true Palladian
feeling. Wren preferred a baldacchino, and indeed left a drawing
for one. So too, it is very difficult to give any approval to the
coloured figures in the spandrils of the dome arches. It is no
approval to say, they are not so bad as they might be. At first
the dean and chapter handed over the cathedral to a gentleman
for whom I must premise that I had the greatest respect and
personal liking. This was the most extraordinary appointment
ever made, in all probability, even by a dean and chapter.
Burges was certainly a man of great taste, but he was an
enthusiastic "Goth"; he saw little or no beauty in St. Paul's,
which he considered in great part "heathenish"; he could not
judge of the beauty of the dome, because he was so blind that
he could not see so far. To frame and carry out a grand scheme
of Palladian decoration, a scheme which was to unite the smallest
and the most distant features in a harmonious whole, in taste
such as we may learn from Vitruvius and the wall-paintings of
Pompeii—this task was entrusted to a gentleman whose whole
mind was centred in the art of the thirteenth century, and who
from physical infirmity was not able to see more than a single
pier of St. Paul's at a time. It is on these accounts that we
tremble for St. Paul's, but there have been signs of late, not so
much of an improvement in taste, as of a recognition of what
style St. Paul's is designed in. The architects of the reredos are
rather halting in their comprehension of the style; but there is

nothing Gothic about it. And an excellent step has lately been taken in the removal of Alfred Stevens's admirable Wellington monument from the baptistery or consistorial court where, at the brilliant suggestion of some bygone dean, it had been so long immured, to a place under an arch of the nave.

By Wren's strength of mind in keeping to himself his full intentions, he enlisted time upon his side. At first the chief thing was to prepare a place for the resumption of cathedral services. He accordingly began with the east end and choir. Ten years were employed upon them, and upon portions of the transepts. Meanwhile, King Charles died, and James succeeded. It was of no consequence now whether Wren adhered to the artificial design or not. James cared nothing except that there should be side aisles and possible chantries and chapels; and, moreover, from the very beginning of his short reign he had only too many things besides St. Paul's to occupy his attention. Service was duly resumed in 1697. The glorious cupola, a modification and improvement upon that in his very first design of all, was not finished for thirteen long years more; namely in 1710, in the reign of Queen Anne.

In concluding these notes, it may be well to answer the only adverse criticism of St. Paul's as a whole which seems to have any weight in it. I say "seems," because it has really no validity, and a moment's examination dissipates it. I may put the objection in the words of those critics who have apparently thought most of it. We do not find any such criticism in Gwilt, Elmes, Godwin, or Miss Phillimore. It is pre-eminently the objection of the critic brought up in the modern mock Gothic school. The objection is that "one half of the building is built to conceal the other half." So I have seen it put, referring to the great wall over the aisle windows, which shuts out of view the buttressing of the middle aisle. Another writer says it is "a falsehood," and actually thinks he could improve it. He suggests that windows should be made into the triforium where Wren made only niches. The obvious answer is that there are windows below the niches and that Wren, for constructional reasons, preferred the solid wall. The Gothic architect would have exposed the buttresses. Wren thought it better, in a Palladian building, to conceal them; and undoubtedly he was right, but even if he had been wrong it would not have been easy to hit upon a better expedient. But suppose, for argument's sake, Wren had consented to expose his buttresses. If we examine a photograph or a careful drawing of the church, we observe that the upper storey of the exterior of the nave consists, commencing at the west end, of a tower of considerable solidity, with a window; of a building with three windows, in which, as we know, on the south side is the library; of three bays, with low windows and the objectionable niches; and finally, of another solid and projecting building, which is admirably placed where it appears to add stability to the dome. From this enumeration, it follows that our Gothic friends would alter the whole character of Wren's nave, for the sake of exposing two single buttresses between the niches just mentioned. This is absurd, and might be so demonstrated by another method: suppose Wren replied that his building would be better here for the weight of a continuous wall rather than of an unsupported flying buttress, there would be no answer. The church has its faults, within and without, but this is not one of them; nor was Wren an architect who wished his buildings to look less stable than they really were.

Numerous as are the drawings for St. Paul's, both by Wren himself and by others his contemporaries, in public and private collections, I believe I am right in saying that no designs exist answering to the church which Wren completed. He made, no doubt, working drawings for his masons; but they have not come down to us. They may have been worn away in the using. He had made many designs for domes, some of which are very like the present dome. On the whole, it seems to be a perfectly tenable view that no complete design was ever made, and that Wren kept to himself what he had resolved upon until it was time to execute it. If we remember the long time — not less than five-and-thirty years — which elapsed between the laying of the first stone and the completion of the whole edifice, this seems to be the most likely view.

I have said something already as to Wren's parish churches, and his objects in designing them. When the Bishop's crusade against them ceases and we come to count our losses, it will be difficult or impossible to reconstruct the subtle harmony which formerly united them. North and east of St. Paul's are the three most beautiful of the stone steeples; and a man must be indeed insensible to the sweet influences of consummate art, if he does not see a meaning in the towers of Christ Church and St. Vedast, leading up to the tower of St. Mary-le-Bow. A very fine tower, resembling that of St. Mary Aldermary, is at St. Michael, Cornhill. The greater part of the church has been much injured of late years by an ignorant "restorer," who thought he could improve upon Wren; but the tower is almost intact. Of elevations, some of the most simple will be found the best. There is nothing else in the City better than the east end in Gracechurch Street, of St. Peter upon Cornhill. A similar design was at St. Olave's, Old Jewry; but the Bishop apparently thought one example enough. Very handsome, but perhaps scarcely as delicate, is the east end of St. Lawrence, in Guildhall Yard. Another, and very plain example, is St. Michael, Wood Street.

All Wren's City churches are worthy of examination; but all are not equally important. In many cases, the unfortunate parishioners could only commission him to spend a trifle on their church. Even in these cases, he always left some pleasing feature—something to make his building, however small and poor, picturesque. In order, however, to study his style accurately, it will be best to confine ourselves to his more ambitious efforts where money was not so scarce, and where he had space in which to display his powers. These larger churches resolve themselves, so far as regards the interior arrangements, into two chief categories. Some, like St. Paul's, have the traditional Gothic plan, with, however, the greatest modifications. Of this class, we may number St. Andrew's, Holborn; St. James's, Piccadilly; St. Bride's, Fleet Street; St. Andrew by the Wardrobe; Christ Church, Newgate Street, and a very few of smaller dimensions. These buildings have galleries in the side aisles, something of a chancel, fine roofs, chiefly of ornamented "barrel vaulting," a conspicuous east window, and many other features, such as we expect in a large ancient church. The gallery is an integral part of the design, not an afterthought, as in most Gothic churches; but it can hardly be asserted by the most partial observer that Wren succeeded from an artistic point of view with his galleries. At St. James's a single Corinthian pillar rises from the gallery, and supports an exceedingly beautiful roof. Below, the gallery is supported by a square prop, of no beauty, and calculated to intercept the view of any one who sits near. At St. Bride's there is a wholly different arrangement. Rising from the level of the pews are coupled Tuscan columns; and the front of the gallery runs along between the columns; a square prop being placed below to increase the appearance, if not the reality, of strength. The pillars support a row of arches, as in a Gothic church, which impart great lightness. Above them, again, are circular windows in the vaulting. At St. James's there are no attic windows; but the church never suffered from any want of light until some modern stained glass was inserted. The Gothic revival, oddly enough, has never taught glaziers that the object of windows is to admit light. In most churches of this character, Wren made distinct provision for a small chancel. In short, had the wishes and schemes of James II. been successful, these churches would have easily been "converted," so as to become convenient for the celebration of Mass. They all partake of the character of St. Katharine Cree, except for their galleries.

It is easy to figure to ourselves Wren turning with pleasure from churches of this model to those in which he could renew his old experiences of the Sheldonian. His conception of a great preaching-house was not fulfilled by St. Paul's or even St. James's. He wanted to roof in as wide a space as possible, without interruptions to the view by pillars or piers. Fergusson reproaches him with being more of an engineer than of an architect; but that shows how little Fergusson understood Wren's work. Whatever happened, he never lost sight of the abstract elements of beauty, such as fitness, picturesqueness, and above all proportion. Some of these plain chambers, with their flat roofs, are full of charm, and give the clearest possible evidence of the intentions of their designer.

The finest of these churches is probably St. Lawrence. It is that in which the great public religious ceremonials of the Corporation take place; and Wren was evidently actuated by a desire to make it in every way suitable to such functions. The seats appropriated to the Corporation are in the centre, and very conspicuous. The church measures 81 feet by 68. The roof, 40 feet high, is ingeniously and beautifully formed of deeply-sunk panels, ornamented with the delicate plaster work which Wren so greatly affected. The central idea of a preaching-house is enhanced by the way in which the roof meets the walls; namely, in a series of coved spaces, with enriched scroll-work, both to give a look of stability not at all necessary, and also with a view to helping the voice without echo or ring. There is no chancel. The holy table stands in its place at the centre of the eastern end. The cost was enormous for Wren—£11,870 : 1 : 9.

Wren employed this method in building one or two other important churches and several smaller ones. St. Vedast, Foster Lane, is a good example, as was St. Mary Somerset, on which the Bishop has laid his destructive hand. Another which has disappeared is All Hallows the Great. St. Mary Abchurch is nearly a square, 65 feet by 60; and St. Michael, Queenhithe, now pulled down, was nearly two squares, being 71 feet by 40. In several of these there is a kind of single side aisle, often so irregular in plan that we should be puzzled, but that we remember Wren's great object was to roof in the whole of the space at his command, and that this space was defined by the old church with its chancel, its aisles, its side chapels, and its chantries; some of them no doubt highly eccentric in axis to the main body of the church. But Wren had in most cases not only to provide for an increased population in the parish, but for the incorporation with it of another parish; and he was obliged, while making but one chamber, to work in every available morsel of space.

In 1679, Wren had to design a church for one of the most irregular of these spaces. This was the site of old "St. Swithin's at London Stone." He showed himself equal to the task, and roofed in every inch of open ground. In doing so, he adopted a new expedient; and St. Swithin's is one of five parish churches with cupolas which he built in the City. The others are, or were, St. Benet Fink, pulled down; St. Mildred's, Bread Street; St. Stephen's, Wallbrook; and St. Mary Abchurch.

St. Stephen's had the good or ill fortune to belong to the Grocers' Company. The consequence is both that the authorities were able to incur a little extra expense in the original design, and also that ever since, with every changing caprice of architectural taste, they have done their best to obliterate Wren's handiwork. The church is very well known; and visitors are fortunate who saw it before the last alteration. I may quote a notice printed in 1823 in Elmes's *Life of Wren*. Elmes forgets grammar, and even sense, in his enthusiasm; but somehow he conveys a very vivid impression of the now departed charm of a building of which Canova said that if he revisited England it would be to see St. Paul's, Somerset House, and St. Stephen's, Wallbrook. Elmes says, "The beauty of the interior of this church arises from its lightness and elegance. On entering from the street, by about a dozen or more of steps, through a vestibule of dubious obscurity, on opening the handsome folding wainscot doors, a halo of dazzling light flashes at

ST. LAWRENCE JEWRY.

once upon the eye ; and a lovely band of Corinthian columns
of beauteous proportions appear in magic mazes before you.
The expansive cupola and supporting arches expand their airy
shapes like gossamer ; and the sweetly proportioned embellished
architrave-cornice, of original lightness and application, completes
the charm. On a second look, the columns slide into complete
order like a band of young and elegant dancers at the close of
a quadrille. Then the pedestals concealed by the elaborate
pewings, which are sculptured into the form of a solid stylobate,
opening up the nave, under the cupola to the great recess which
contains the altar, and West's fine historical picture of the stoning
of St. Stephen, lift up the entire column to the level of the eye ;
their brown and brawny solids supporting the delicate white
forms of the entire order."

The last part of this curious passage—that relating to the
pewings—should be specially noted. The arrangement of the
dark oak wainscoting produced a most interesting scenic effect.
When you entered from below, the church seemed to rise above
you. All its architectural features began to show, so to speak,
above the level of the tall sombre pews. The size, and especially
the height of the church were so enhanced that it was impossible
to believe that it was only 87 feet 10 inches by 64 feet 10 inches,
with 63 feet to the top of the highest part of the dome.
Fergusson, who was no enthusiastic admirer of Wren, says that
here he produced " the most pleasing interior of any Renaissance
church which has yet been erected." Further on he repeats :
" There is a cheerfulness, an elegance, and appropriateness
about the interior which pleases every one." The leading idea
of the architect was to place " a circular dome on an octagonal
base, supported by eight pillars," and Mr. Fergusson considered

this was an "early and long a favourite mode of roofing in the East, and the consequent variety obtained by making the diverging aisles respectively in the ratio of 7 to 10, infinitely more pleasing than the Gothic plan of doubling them, unless the height was doubled at the same time." What Fergusson meant by "the East," I do not know. There was nothing to compare to St. Stephen's in India, Syria, or Egypt before the time of Wren, whose design, in any case, must be accounted wholly original.

This church has always laboured under the same disadvantage as St. Paul's. The authorities concerned with it have always had too much money. I have not heard that St. Stephen's has been scheduled for destruction by the committee; but after the "restoration," we may regard its ruin with comparative equanimity. The great scenic charm of the interior has been carefully and elaborately removed. It no longer bursts upon the view as we ascend from what Elmes calls "the vestibule of dubious obscurity." The interior has been gutted. The panelling which had such a magic effect has been removed. The floor has been laid down with coarse mosaic. The pedestals of the pillars are exposed, with a disastrous result; and in the centre a few yellow oak seats, fresh from Tottenham Court Road, have been placed, as if to accentuate the smallness of the congregation. We all admire courage, and perhaps some readers would like to know the name of the gentleman who ventured so boldly to improve upon Wren's masterpiece. It is Peebles, and he is understood to be a very accomplished architect.

Mr. Wheatley says that Wren was averse to the use of these panellings, and that they were forced upon him by the Grocers'

Company; and Miss Phillimore speaks of "the disfiguring pews" which she desired to see removed. Neither of these writers apparently understood that even if they were forced upon Wren, which I must take leave to doubt, he used them in such a way as to make them an integral part of the design. Tinkering of all kinds has gone on for many years, and the "restoration" of Mr. Peebles was only the final step in a long series of such ruinous operations. Among the first was a frightful vandalism, the insertion of mock-mediaeval stained glass in the windows. But the treatment of this little gem of architecture is not a subject pleasant enough to be dwelt on here. It has always been very difficult to obtain access to the interior on a week day; and the visitor need not now go to the trouble which in Canova's time and later was necessary before the key could be found.

It has often been remarked by architectural writers that St. Stephen's would form an admirable model for a modern church. Several attempts in this direction have resulted in failure. The reason is easily found. If an imitator either enlarged or diminished St. Stephen's, the proportions would be lost. A St. Stephen's double the size would have a wholly different effect. It is so small that the imitators have generally tried to build something larger; but there would be great difficulty in making the needful calculation. It cannot be done by rule of thumb. It may be worth while here to mention that some admirable drawings of St. Stephen's, by Mr. Edmund H. Sedding, were engraved in the *Builder* on 3rd January 1885, having gained the Royal Academy medal in 1884. The drawings were made before the church was "restored."

St. Mildred, Bread Street, was finished in 1683. It is but

small, 62 feet by 36, but obtains a certain dignity from the dome, which rises to a height of 40 feet. The plaster ornamentation is even worse than that in the dome of St. Stephen, Wallbrook, but the exquisite carving of the pulpit and of the fine Corinthian reredos goes far to redeem it. This church deserves more attention than it usually receives, both as an example of how to make the most of a very small site and of how to build a cupola in the simplest manner and with the most ordinary materials. It is technically described as follows :—It is formed within the external roof by means of slight deal ribs attached to the principal timbers. They are lathed and plastered. The whole roof is of an ordinary tie-beam and king-post construction; but in that part which occurs immediately over the cupola, the tie-beam, instead of being attached to the foot of each principal rafter, as usual, is raised about half-way up, in order to admit the rise of the cupola; and diagonal braces from rafter to rafter are introduced. I take this description from Godwin and Britton, who add, "the architectural student may derive advantage from an examination of it."

St. Mary Abchurch is built on a slightly more ambitious scale, as it is almost a square, being 65 feet by 60. It was finished in 1685, all but twenty years after the fire. It is well worth a visit, as well for the ingenuity of the plan, for the beauty of the design, and for the exquisite finish of the carved oak, in this case, there is little doubt, by Grinling Gibbons himself. The cupola was decorated with painted angels by Sir James Thornhill, which look far better than the plaster angels in St. Mildred's. The altar-piece, also of Corinthian columns, is particularly fine. The gallery, added in 1822, rather accentuates than conceals the irregularity

of the plan. The cost of this handsome church was only
£4900.

St. Swithin's resembles St. Mary Abchurch in several
particulars, but the exterior is much more ornamental. It is
61 feet long, 42 feet from east to west, and 40 feet high. The
roof is formed into an octagon cupola by a very ingenious
arrangement of the timbers. The rest of the church has been
much pulled about by "restorers," one of whom, in 1869, made
a determined attempt to Gothicise it.

Of Wren's Gothic churches it is not needful to say much.
The great spire of Lichfield Cathedral has lately, on questionable grounds, been attributed to him. The most interesting
in London is St. Dunstan's in the East, the tower and spire
of which will remind the spectator of St. Nicholas at Newcastle; but the mouldings and other features in their delicacy
are Wren's own. I know that many competent judges do
not admire Wren's Gothic mouldings; but at least they are
better than their coarse substitutes in modern Gothic. Both
may be studied in St. Michael upon Cornhill. St. Dunstan's was lately condemned, incredible as it may seem.
The parishioners were amused with the usual tale. The
steeple, it was said, had been wholly renewed since Wren's
time, and the design altered. A provisional consent was
wrung from them; and the committee gave out that the
church was about to be destroyed. Fortunately, however, the matter came to the ears of a gentleman who had
been connected with St. Antholin's, Watling Street; and he
recognised the same old story that had worked so effectually
with him and his fellow-parishioners in 1875. A letter in the
Times betrayed the whole scheme. The provisional vote

was at once cancelled. The church is still, of course, in imminent danger. The agent will next time, no doubt, have invented a fresh assertion; and as long as these precious buildings are at the mercy of parishioners who can be cajoled or coerced by interested agents, we are bound to feel uneasy. Another Gothic church by Wren was interesting; that of St. Alban, Wood Street. It has been bedizened by a modern mock Gothic architect, and all its points of interest removed. Too much has also been done to improve Wren at St. Mary Aldermary; but it is still in a fairly genuine condition.

VIII

THE SUCCESSORS OF WREN

VIII

THE SUCCESSORS OF WREN

Vanbrugh—Hawksmoor—Gibbs—James—Archer—Burlington—Campbell—Kent—Taylor—Chambers—Adam—Wood of Bath—Baldwin—Palladian in the Provinces—Dublin—The Bank—The Four Courts—The Custom House—Trinity College—Barry's Club Houses—The Grecian Style—The Reign of Stucco—The New Gothic—Conclusion.

In one respect, Wren fared better than Inigo Jones. He both founded a school and lived to see it flourish. Though he had practically retired many years before his death in 1723, he could see around him several architects well able to take up and carry on his tradition. The best of them was undoubtedly Burlington, who had already commenced operations in Piccadilly; but of Wren's immediate followers, Hawksmoor who was his pupil, Gibbs who worked on parallel lines, Archer, James, and two or three more were all in practice. Vanbrugh, however, was the only one who showed much originality; and this is perhaps as well. Outside the pale of his immediate following were Kent, whose work singularly resembles Wren's, and can, in fact, hardly be distinguished from it either at Kensington or Hampton Court; and Colen Campbell, both of whom worked under Burlington. From him they derived anything of the nature of originality or genius which either of them showed. In addition to all these, we must not overlook the claims of the elder Wood, who, in 1728, five years after Wren's death, began the north side of Queen's Square at Bath—assuredly the most satisfactory piece of town architecture erected in the style up to that date.

We need not dwell long on the career of Vanbrugh. He resembled certain very successful architects of our own day. If he had any knowledge or any artistic perceptions, he carefully concealed them. His chief country houses, Castle Howard and Blenheim, amply justify the mocking epitaph :

> Lie heavy on him, earth, for he
> Laid many a heavy load on thee.

His style consisted in the negation of style. If a Composite column should be no more than ten diameters high, Vanbrugh made it twelve or fourteen. At Castle Howard, he treats us to two rows of columns on the same front, both of the same Corinthian order, yet the proportions of the two rows are different. At Blenheim his columns are few and unobtrusive ; but his windows with their thick ungraceful mouldings have the most gloomy effect. Similar windows occur also at Greenwich, and seem in fact to have been a necessary part of most of his buildings. As for proportion, he cannot have thought much about it ; but if he did, he seems to have been satisfied when he made a house look like a fortress, or a palace look like a mausoleum. Nevertheless he was in vogue, partly, no doubt, as a kind of reaction against the strictness of the Palladian rule. He talked cleverly. He wrote plays. He posed as a wit and a man of fashion ; but so far as my limited powers of observation go, his genius, allowing that he had any genius, was the reverse of Wren's. Wren had a capacity for taking trouble, Vanbrugh a nearly equal capacity for saving himself trouble. I cannot account for it that Fergusson, who

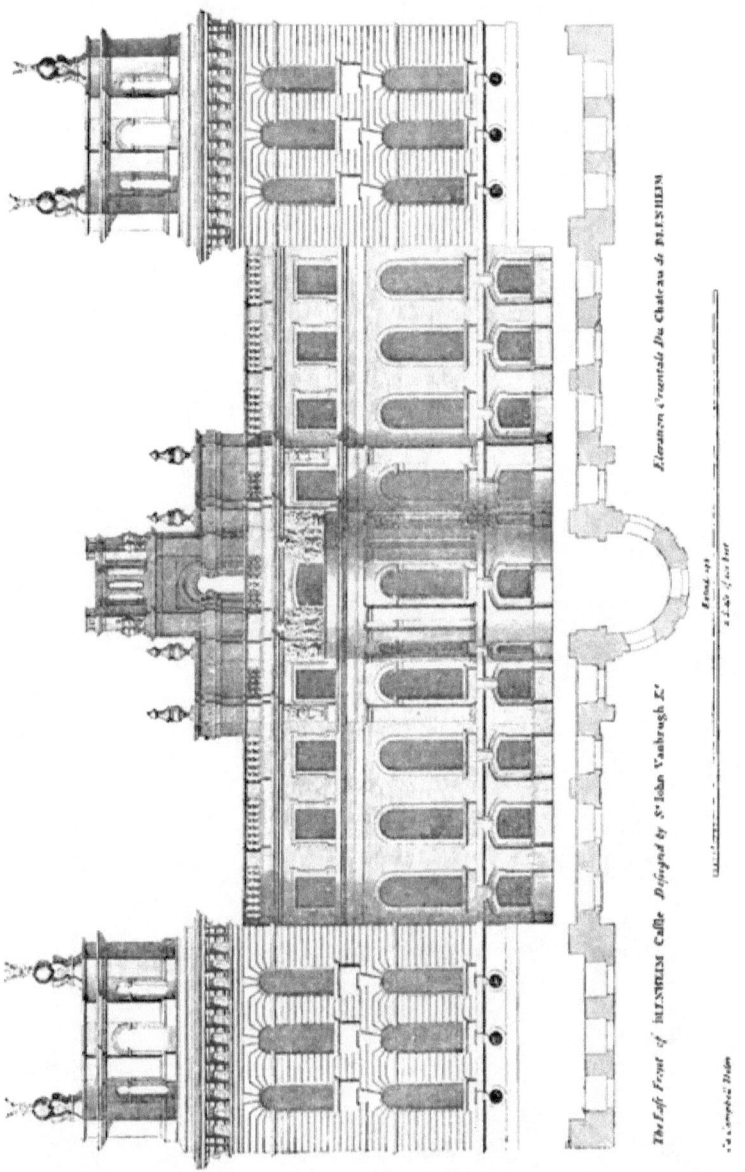

is severely critical of Wren, has nothing but praise for Vanbrugh. He says of him (p. 283), " He never faltered in his career; and from first to last—at Blenheim and Castle Howard, as at Seaton Delaval and Grimsthorpe—there is one principle runs through all his designs, and it was a worthy one—a lofty aspiration after grandeur and eternity." Though the English of this passage is questionable, the meaning is perfectly clear, and I can only suppose that Fergusson could understand the crude art of Vanbrugh, and could not appreciate the subtlety of Wren, whom, indeed, in one passage he characterises as more an engineer than an architect, a singularly unfortunate remark. True, Wren was an engineer and an astronomer, and much besides; but though in these pursuits he has since been often excelled, as an architect he still stands alone. Vanbrugh was constantly set up in opposition to Wren, and is said to have been commissioned to design Blenheim, because Wren had offended the duchess about Marlborough House in Pall Mall; but the sequence of the two events is reversed by some authorities. Vanbrugh was chosen, whatever the reason; and it was an unlucky choice. Blenheim, as we see it, is a monster of shapelessness and, indeed, positive ugliness, and does not contain a single exterior feature, or a single interior apartment, of which we can speak in praise. Everything is enormous in scale and anomalous in design. Fortunately, Vanbrugh was not employed to build any important churches; and, in fact, his style, or want of style, is now chiefly interesting as marking a reaction against Wren.

The rules were very strictly observed in the works of Hawksmoor, Gibbs, James, and others, who found in them a refuge from the absolute want of originality which characterised them all. Hawksmoor was employed as clerk of the works both to Wren and to Vanbrugh during a considerable part of his career. The passing of the act for building fifty new churches brought him some tasks of a public character; and the best of the fifty is undoubtedly his, namely St. George's Bloomsbury. An eminent Gothic architect was turned loose upon it lately, with the usual effect. He removed the galleries of which Hawksmoor was probably most proud; but he could not destroy the noble portico of Corinthian columns. He took down the characteristic lion and unicorn from the steeple, but did not take down the steeple itself, said to be an attempt to realise ancient descriptions of the mausoleum at Halicarnassus. A few years ago, the temporary removal of some surrounding houses enabled us to see the extraordinary picturesqueness of the body of the church. Hawksmoor did not reach the high level of St. George's again. His St. Mary Wolnoth is too heavy, though picturesque; and in the interior the gallery question, which so greatly taxed even Wren, came very near to a solution. His St. George's in the East acknowledged the power of Vanbrugh's influence, and, though lighter and more graceful than anything of Vanbrugh's, is quite as anomalous. The towers at All Souls at Oxford, like so much of his work, are picturesque; and altogether it must be said of Hawksmoor that, more than any other of Wren's immediate successors, he showed signs of being able to think for himself. Born in 1661 he survived his great master thirteen years, dying in 1736. James Gibbs, whose churches in London are much more conspicuous, was twenty years younger, and lived till 1754.

Gibbs has left an interesting volume of engravings from his own designs. His anxiety to attain or choose the best sometimes amounts to fastidiousness and sometimes degenerates into irresolution. It never resembles Wren's endless " capacity for taking pains," nor, on the other hand, is it like Vanbrugh's lazy striving for originality. Gibbs habitually committed one serious crime in designing churches. He placed the tower on the top of the portico. When we look at St. Martin's in Trafalgar Square, our admiration is constantly harassed by the thought of how much finer it would have been if the tower had been north of the portico. The view from the south side of the square would then have been the finest of its kind in London. But Gibbs missed it, and has more than half spoilt his grand Corinthian portico by imposing his tower and spire as if they grew from the roof "without any visible means of support." Wren always set the base of his steeple squarely on the ground, and seldom set it exactly at one end or other of his church. Gibbs transgressed again in the same way, but less flagrantly, in St. Mary-le-Strand, as well as in a design for a round church, in his book, never carried out.

Some proposals have recently been made to alter the portico of St. Martin. The exact object of these proposals has not transpired. The roadway is more than wide enough for the traffic; and if any building in the neighbourhood is to suffer, let it be the National Gallery on the opposite side. The portico, in architectural language, is hexastyle, of the Corinthian order; the intercolumniations being of two diameters and a half, and the projection of the portico of two intercolumniations.

Conspicuous as St. Martin's is, the little church of St. Mary-le-Strand is nearly as well known; and if a projected improvement be carried out it will form the central feature of a new square. Gibbs built it seven years at least before St. Martin's; and it cannot be said that his art improved in the interval. The proportions of St. Mary's are exceedingly delicate, and were evidently thought out by the architect with a view to adorning the situation. A proposal to remove the whole church and rebuild it elsewhere was gravely made three or four years ago; but St. Mary's would simply look absurd in any other place than the narrow and crowded thoroughfare for which it was designed.

Gibbs's other designs call for little notice. He built a considerable part of St. Bartholomew's Hospital. He designed the Senate House at Cambridge for Sir James Borough: part of a larger plan never carried out. Sudbrook, near Richmond, is still very much what it was while it was inhabited by the great Duke of Argyll and Greenwich; but none of these buildings call for special remark. Of his churches, St. Peter's in Vere Street has been thoroughly refitted within the past few years; and perhaps the handiwork of Gibbs will be most easily recognised at Whitchurch, near Edgeware, which he rebuilt for the Duke of Chandos. The east end is lined with carved oak, and contains an alcove of the Corinthian order, in which is placed the organ of the immortal Handel. The Radcliffe Library at Oxford is perhaps the most important of Gibbs's works, and is well known for its delicate proportions and graceful dome, and for the charming feature it makes in every view of Oxford.

One of his most pleasing compositions is a monument in the north transept of Westminster Abbey. It was erected by "the Lady Henrietta Cavendishe Holles Harley," afterwards Lady Oxford, to the memory of her father, the Duke of Newcastle. There seems to have been what is now common, a competition among the architects for this monument, as Gibbs says in his book, already mentioned, that, "this draught was pitched upon

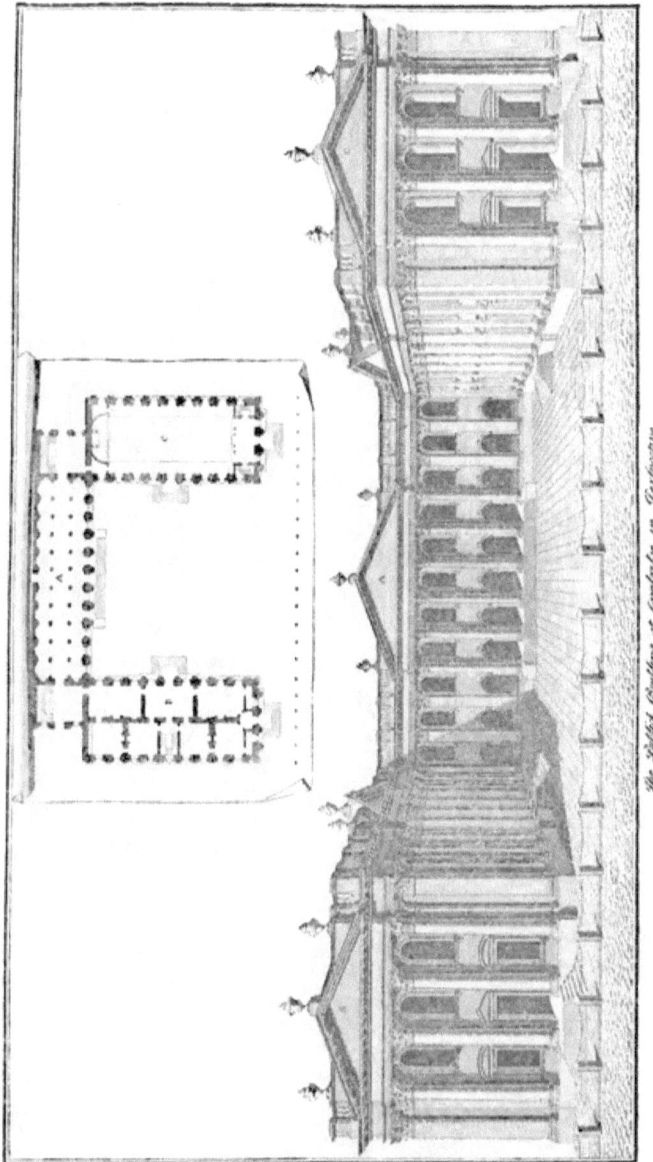

The Publick Building at Cambridge in Perspective
A The Royal Library
B The Consistory of Regents House
C The Senate House

BY JAMES GIBBS.

ST. MARY-LE-STRAND, BY GIBBS.

amongst many others." The print in Gibbs's book is by Vertue, but there is another by Cole, in Dart's *Westminster Abbey*, which represents the design as it is, with the front curved. When it was new, with its gilding and heraldry, and the coloured marble Corinthian columns, this monument must have been very handsome, especially as there were hardly any others at that time in the transept, now so crowded. Gibbs also designed for Westminster Abbey monuments to Prior the poet, and Ben Jonson, both commissioned by Lord Oxford, and to a Mrs. Bovey and a Mr. Smith. Lord Oxford furthermore employed him to build a church in Vere Street, a very plain performance. Gibbs, who came from Aberdeen, was, like Inigo Jones, a Roman Catholic, but was buried in what used to be then the church of St. Mary-le-Bone, now the parish chapel, and near him Rysbrack, the sculptor, who had carried out so many of his designs.

Of James, there is little to be said, except that he built St. George's, Hanover Square, a heavy but handsome church, and St. Alphage at Greenwich, which is an unsuccessful attempt to attain picturesqueness by eccentricity. It is sometimes attributed to Hawksmoor, but looks much more like the work of James. Another architect of this school was Archer, who built St. John's church at Westminster, which has been unsparingly criticised, chiefly on account of its four corner towers, or belfries, as Walpole calls them. But the objectors did not notice that these towers were a structural necessity, as when the church was built it was found not to be strong enough for its situation in marshy ground. Archer also designed Cliefden House, which has been much altered of late. We should also mention Vardy, whose Spencer House, in the Green Park, is modelled closely after Inigo Jones. It is figured in the *Vitruvius Britannicus*, iv. 37. He also designed Uxbridge House, Burlington Gardens.

The greatest of all the architects who followed Wren in the first half of the eighteenth century was, strange to say, strictly speaking, an amateur. This was Richard, third earl of Burlington and fourth earl of Cork, who was born in 1693, and died in 1753. If Burlington had enjoyed the good fortune to be born poor, his fame as an architect might haply have rivalled that of Inigo Jones, if not that of Wren himself. Burlington had an unbounded admiration for Inigo and all his works. It was by his superintending munificence that Kent was able to publish the two beautiful volumes of Jones's designs; and, as is well known, he restored and preserved the church in Covent Garden. His modesty exceeded even his ability, and he willingly permitted so inferior an architect as Colen Campbell to claim and receive the praises earned by the beautiful design of Burlington House. Unfortunately, very little of his work can be positively identified. In London, Burlington House has been practically destroyed. The house of Marshal Wade has disappeared, though it survives in the courtyard of a hotel, and still has its lovely staircase. The design is also preserved in the Provost's House in Dublin, imitated by an architect named Smith. The best specimen of Burlington's work that exists is hidden away behind Westminster School. A small house with wings, at the end of Savile Row, was built by him in his garden. In the country we have the villa at Chiswick, slightly altered, and much added to, but still in such a condition that we can judge of its merits. At Bath, a house in the Orange Grove is attributed to him; but I cannot recognise his touch in it. At York, where he was Lord Lieutenant, he designed the beautiful Assembly Rooms. It is a remarkable fact that we should not be able to distinguish his designs from Inigo Jones's in Kent's volumes if they were not all signed. There is certainly nothing in English architecture more quietly beautiful than the elevation of the Westminster dormitory. (It is plate 51 in the second volume of the book of Inigo Jones's designs.) Fifteen arches support the upper storey, which contains a single chamber, 166 feet long. The whole building is 55 feet high. Above the arcade is a row of niches for statues, and above that again, a row of small square window apertures. At the time when the dormitory was built, the site, which looked on the College Garden, was very damp, hence the elevation of the chamber itself on arches. Now that the ground is thoroughly drained and dry, the arches have been built up and further school accommodation has been obtained; but this alteration has been carefully made so as to interfere little with the effect.

Campbell (*Vitruvius Britannicus*, iii. p. 21) takes to himself the credit of having designed Burlington House; and it is more than likely that he made all the working drawings. It is, however, equally certain, first, that he never was at Vicenza, where is the building by Palladio from which the design is said to have been taken; and, secondly, that no recognised building of his—Wanstead, or Houghton, or Mereworth—is good enough to be by the same designer as Burlington House. Burlington had been much in Italy, and had seriously studied the art of Palladio. The Chiswick villa was also an adaptation after the same great architect. A second point in the argument against Campbell's authorship is, that his claim was not acknowledged at the time. Walpole, in particular, treats it with contempt,

while he goes into raptures over the colonnade. In the next
age, Sir William Chambers, the best possible judge, considered
the whole composition to be unrivalled, and, in his work on
Civil Architecture, makes use of the oft-quoted phrase, " Behind
an old brick wall in Piccadilly there is, notwithstanding its
faults, one of the finest pieces of architecture in Europe." It
is only wearisome to go on quoting what Ralph, Pope, Gay,
Malcolm, and Britton say in praise of this beautiful building.
It is better worth while to put a few facts relating to it into
chronological order. In 1716, Burlington, not yet of age, met
Kent in Italy. He had already, in 1715, been appointed Vice-
admiral of York, an Irish privy councillor, and colonel of a
militia regiment. In the following year, 12th January 1716,
he was made Governor of the County of Cork. Four months
later, he was appointed Lord Lieutenant of the West Riding
and City of York. It is evident that at twenty-one, when he
returned home from Italy, full of what he had seen and of the
friendship he had already formed with the young architect he
left behind, he had quite enough to employ his mind both in
public and in private business. He, Burlington, doubtless made
careful drawings of buildings at Vicenza; and, as soon as he
could secure the services of an architect, having probably no
time of his own to spare, he set about refronting the old house
in Piccadilly in the Italian style. He must have begun as
early as 1716, for that date with his arms was on the leadwork;
but Campbell specially dates his view of the house 1717, and
that of the gate 1718. Whether after this time Lord Burlington
had further dealings with Campbell we do not know. Campbell
was in some way mixed up in the intrigue by which Wren was
ousted from office, and died the same year that Kent came

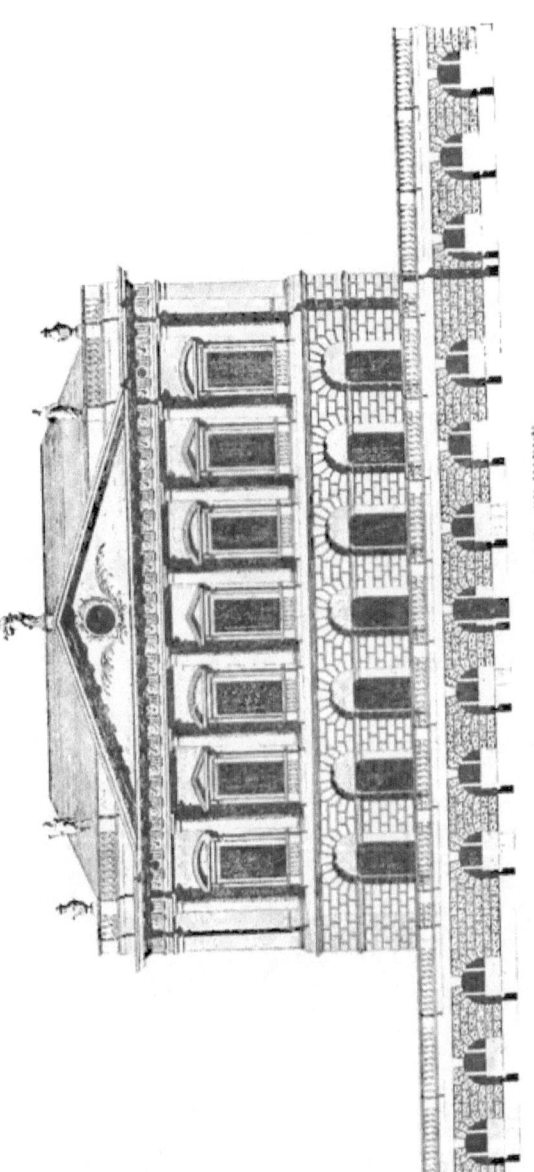

SPENCER HOUSE, GREEN PARK. BY VARDY.

home, 1729. Kent was welcomed to Burlington House by his friend. So warm was this welcome, and so great the friendship of the pair, that Kent never left Lord Burlington again, but, having lived here for nineteen years, died in the house and was buried in the vault of the Boyle family at Chiswick in 1748. Meanwhile, the two published the great book upon Inigo Jones I have so often had occasion to mention; and we can imagine with what pleasure Kent included a few designs by his friend and host with those of the great architect. Kent did not leave very much mark on the public buildings of London. He is believed to have added to Kensington Palace. A state staircase there is always attributed to him, and he decorated the principal apartments. The cupola room, if it is his and not Wren's, does him great credit. Even in a dismantled state, it is fine and rich. The doorways and niches, and the fireplace, all of the Ionic order, in marble, produce a magnificent effect. Kent is certainly responsible for the painting of the staircase, which he carried out, with figures intended to enhance the appearance of size. The flat ceiling is painted to represent a dome, and faces peer down through the skylight. He was also employed at Hampton Court to complete some of the work Wren had begun. It is not always possible to distinguish his work, but there cannot be much doubt as to a chimney-piece in the Queen's Guard Chamber, which is supported by figures of guards. Holkham Hall is usually considered Kent's best domestic work, the influence of Lord Burlington's greater taste and originality being very plainly marked upon it. Sir William Chambers, however, finds fault with it. In the north front (which is represented in our plate from the *Vitruvius Britannicus*,

v. 25), there are no less than seven Venetian windows, "which, added to the quantity of trifling breaks and ups and downs in the elevation, keep the spectator's eye in a perpetual dance to discover the outlines, than which nothing can be more unpleasing or destructive of effect." The south front is more satisfactory, as it is set off by a magnificent hexastyle portico of the Corinthian order, raised on a rusticated basement. Chambers also makes some disparaging remarks about the Horse Guards, but there Kent was associated with Vardy in the design. The old part of the Treasury buildings was erected in 1733 by Kent, part of a much larger design, never completed. It is plain but perfectly satisfactory as regards proportion and features.

During all the years of their association Kent and Lord Burlington seem to have very seldom worked together, and to have retained, except in the examples I have named, a complete independence of style. But Burlington House was designed and finished before Kent came from Italy.

Mr. Wheatley and others say it was designed in imitation of the palace of Count Valerio Chiericato at Vicenza. Burlington did make a design after this palace (*Inigo Jones*, ii. 12), but it was never used. Ware, in his *Palladio*, gives two engravings of the Chiericato palace. Burlington House did not in any way resemble them, nor did it, strictly speaking, exactly resemble any of the numerous villas at Vicenza, which Ware has engraved. Part of the house of Count Ottavio de Thieni would give us a hint of the general motive of the design, with its rusticated lower storey and the columns above. But the columns are of the Composite order, whereas Burlington's are Ionic. The following is a technical description

HOUSE DESIGNED BY LORD BURLINGTON FOR GENERAL WADE, IN BURLINGTON STREET.

taken from Britton and Pugin: The south front is in three divisions with a rusticated basement; the central, with six windows, being recessed from the two ends. The first storey, or principal suite of apartments, is ornamented with six columns in the middle division, and four pilasters in the front of each end. In these ends we find the Venetian windows have, very judiciously, been raised to range with the seven other windows. This storey is crowned with an appropriate entablature and balustrade.

It is just possible at Burlington House, as we now see it, for the judicious visitor to make out a glimpse of the upper storey. A third storey has been added, and below there is a kind of portico, both not only incongruous but thoroughly bad in themselves. The unhappy architects, finding they had wholly failed to hit off the proportions or any of the feeling of the original, covered the wretchedness of their design with a wealth of ornament which only serves to enhance its deficiencies. The famous colonnade has, I hear, disappeared; but the gate, with the stones numbered, lies in the mud at the western entrance of Battersea Park, where it forms a kind of gymnasium or playground for swarms of children. In its place stands a new entrance, which, were it not for its enormous height, might escape observation for its architectural insignificance. In spite of a lavish use of ornamental carving, it may fairly claim to be, within two at the most, the ugliest building in Piccadilly.

The York assembly rooms have been added to, and were subjected some years ago to a redecoration by Owen Jones; but the following notes, taken from the volume in the *Beauties of England and Wales* on Yorkshire, tell us what they were like in 1812: " The magnificent Assembly Rooms, erected in the last century, and designed by the Earl of Burlington, are an honour to the city and to the architectural taste of that nobleman. The grand room is an antique Egyptian hall from Palladio, 112 feet in length, 40 feet in breadth, and 40 in height. This room consists of two orders: the lower part, with forty-four columns and capitals and a beautiful cornice, displays the Corinthian order; the upper part is after the Composite, richly adorned with festoons resembling oak leaves and acorns, with a superb cornice, curiously ornamented with carved work." There are smaller rooms: one 66 feet long, another a cube of 21, and a circular card-room. The front had a semicircular portico, but I fear it has lately been removed, or "restored." There are several views and a plan in the *Vitruvius Britannicus*, iv. 78.

Of Chiswick, it is only necessary to say that it was a mere summer-house, but a very pretty one. Two wings had to be built to make it habitable. The original design closely resembles that of the villa near Vicenza of Monsignor Almerico (p. 13 in *Ware*), which was designed by Palladio; but there are many differences, and, while the order at Vicenza is Ionic, that at Chiswick is Corinthian. There are three plates representing it in Kent's *Inigo Jones* (vol. i.). Walpole set the idea going that the design was from that of the Villa Capra. The Villa Capra in Ware's *Palladio* is wholly different; but a circular villa, very like that to which he gives the name of Almerico is still at Vicenza, and has " Marius Capra Gabrielis F " over the portico. It is figured in Schütz (*Die Renaissance in Italien*, Abtheilung B.), and resembles both Ware's Almerico villa and this one at Chiswick. It is evident that there is either some confusion here, or that Chiswick is wholly original, and that Horace Walpole was not the most accurate of authors.

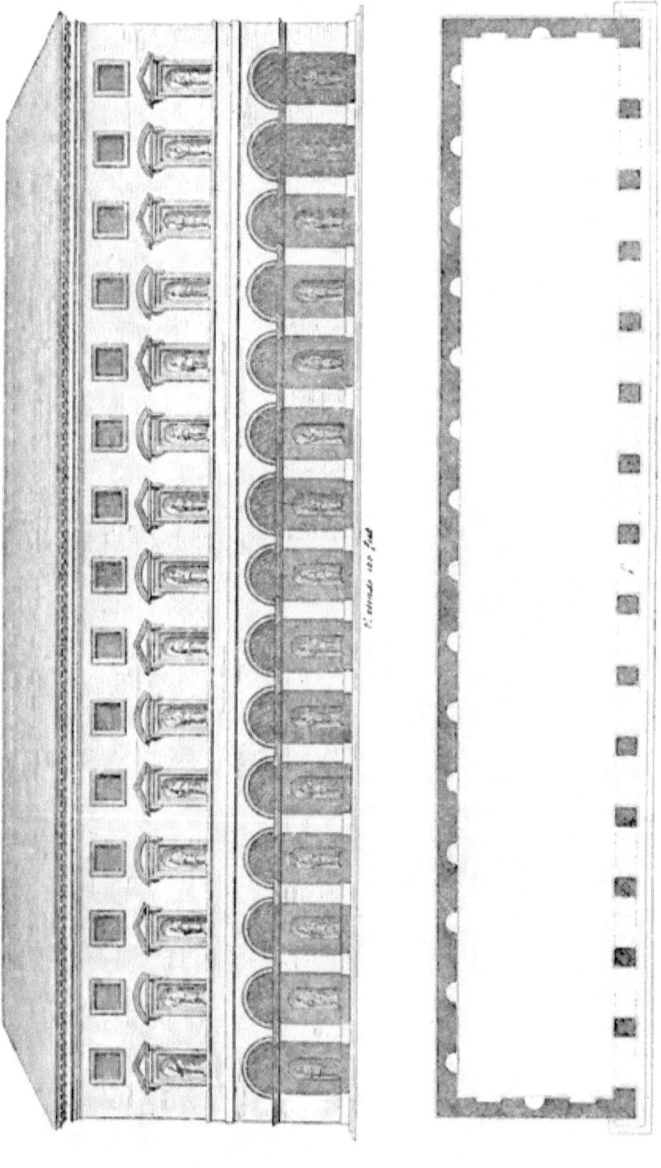

DORMITORY: WESTMINSTER SCHOOL. BY BURLINGTON.

We have so little of Lord Burlington's, that it is perhaps rash to praise him very highly. If he had been an architect in ordinary practice it is possible, nay, probable, that his art would not have maintained itself at the high level at which alone we see it. Kirby Hall in Yorkshire (*Vitruvius Britannicus*, v. 71), though designed by Burlington, was carried out by Morris, one of his architectural followers, and Harewood (*Vitruvius Britannicus*, v. 25), which has many traces of his hand upon it, was by Carr. But enough remains of what is undoubtedly his to justify us in ranking Burlington very little below Inigo Jones.

It would be easy to fill up the rest of this chapter with a criticism of the works of the succeeding professors of Palladian architecture. Taylor was the eldest, having been born in 1714. He had a large practice, but left little that calls for remark. He designed the stone building in Lincoln's Inn, and, it is said, proposed to pull down all the other buildings and lay out the whole inn afresh. James Paine built a good many houses in a pleasing style, but is now little remembered. Chambers is chiefly remembered as the architect of Somerset House, of which the Strand front is always considered a transcript of a former building close to the same site by Inigo Jones. It is probable, to judge by pictures and prints, that Chambers made some such attempt at imitation. But if we remember that Jones's Somerset House faced south, that it was of brick with stone dressings, and, especially, that it was much smaller, we can understand that no such attempt could be successful. The difference of size, material, and aspect would be fatal to any possibility of an exact copy. The front is, however, though unlike Inigo, undoubtedly very fine; and the arched entrance is one of the most graceful examples of the kind in

London. The western side in Wellington Street is also good; but it is impossible not to agree with Fergusson when he says that the south front was Chambers's great opportunity, and unfortunately shows "how little he was equal to the task he had undertaken." His happiest efforts were on a much smaller scale. His arbours, and alcoves, and summer-houses in Kew Gardens, chiefly designed for the Princess Dowager of Wales, are one and all extremely pretty. He published a very charming volume about them. Not far off, at Rochampton, there is a kind of temple of the Composite order which will bear a great deal of examination. It and the villa to which it belonged were designed for Lord Bessborough. The villa was lately pulled down. He built in London Lord Gower's house in Whitehall, Lord Melbourne's house in Piccadilly, and the Albany. He was born at Stockholm in 1726, and afterwards went to China, where he made drawings of Chinese buildings, having, even in his youth, a strong taste for architecture. He rose in this art to the highest honours, was the first Treasurer of the Royal Academy, Surveyor-General of royal buildings, and architectural tutor to George III., who is said to have himself, no doubt with his teacher's help, designed a house, the Ranger's Lodge in the Green Park, now pulled down. Chambers practically retired from the profession some time before his death in 1796. Next to him, in date, comes Robert Adam, born in 1728. Adam and his brothers obtained a great reputation for a knowledge of architecture; and their book on Spalatro enhanced it. In London, their most conspicuous building gives no just idea of their powers. The Adelphi Terrace served for many years as a foil to Somerset House, in which capacity it has now been superseded by two or three new

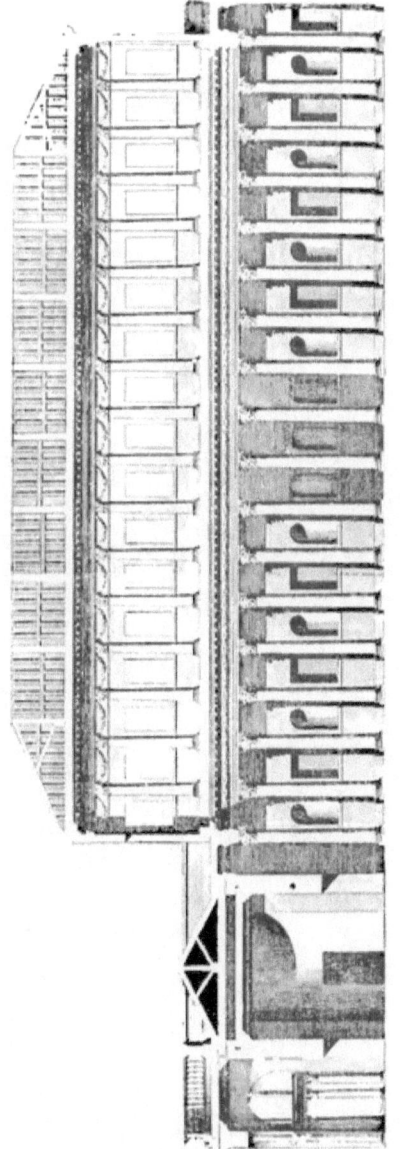

ASSEMBLY ROOMS, YORK. BY BURLINGTON.

blocks of surpassing ugliness. Lansdowne House is very tame. But Adam did better things, as for instance two sides of Fitzroy Square. Kedleston Hall in Derbyshire has many merits; the design, and especially the singular arrangement of the plan, with its four wings, being an adaptation from Palladio. (Ware's *Palladio*, B. ii. 58.) There are several views of Kedleston in the fourth volume of *Vitruvius Britannicus*.

In London, the taste of the next generation was wholly taken up with what was supposed to be Grecian. This name was gradually applied to everything that was not Gothic. One writer calls the towers of Westminster Abbey "Grecian." In a modified degree, some of Soane's work may be called Grecian; but the beautiful adaptation of a temple at Tivoli, which he placed at the north-west corner of the Bank of England, is purely Roman. The greater part of the bank was built by Sir Robert Taylor before 1788, but Soane remodelled the whole. The exterior, with the exception of the corner just mentioned, is studiously plain, though handsome. The courts are more ornamental, one of them, by the way, having been formerly the churchyard of St. Christopher. This one is by Taylor, but the beautiful "Lothbury Court" is by Soane. Leeds describes it as highly picturesque, which it certainly is, and adds "it is not easy to conceive a more beautiful composition." Technically he thus describes it :—" On either side is a flight of steps, the entire width of the court, on which rest two beautiful colonnades of four Corinthian columns, with antæ and entablature ; that on the right hand forms an open screen to a raised part of the court, and that on the left a loggia, the centre part of which is a large semicircular recess, extending the width of three intercolumns." By some chance, not easily accounted for, Soane, forgetting what he had learned in Italy, took to designing in an anomalous style, and his law courts at Westminster, which have now disappeared, were as plain and uninteresting as anything in London.

Meanwhile, true Palladian was flourishing elsewhere. At Bath, where excellent building stone was to be had easily, the two Woods, father and son, and Baldwin the city architect were at work. Queen's Square, built in 1729, I have already mentioned. It is remarkable as apparently the first example of a design which grouped a row of separate dwelling-houses into one composition. The same idea was in Adam's mind in building Fitzroy Square; but Queen's Square was built when he was only a year old. At Bath, the Woods carried it much further; in Great Pulteney Street, perhaps, too far. But no one can help admiring the Circus, which groups together thirty handsome residences, with a spacious enclosure of grass and trees before them. The houses are built with columns engaged and coupled, in three storeys. The lowest is of Tuscan Doric, Wren's favourite order. The entablature is carved with emblems, apparently from one of the emblem books which were so common two hundred years ago. The series is continued, without repetitions, round all the houses. The columns of the next storey are Ionic, beautifully carved, but they have suffered much from weather and age. At the top the order is Corinthian. The whole arrangement resembles the cloister of St. Giustina at Padua by Palladio, but is thus summed up in one of the local guide-books: it "is a fine circle of houses, divided into three blocks, which consist of three stages, each *built in a different style of architecture.*"

The Royal Crescent faces south and has a beautiful view of

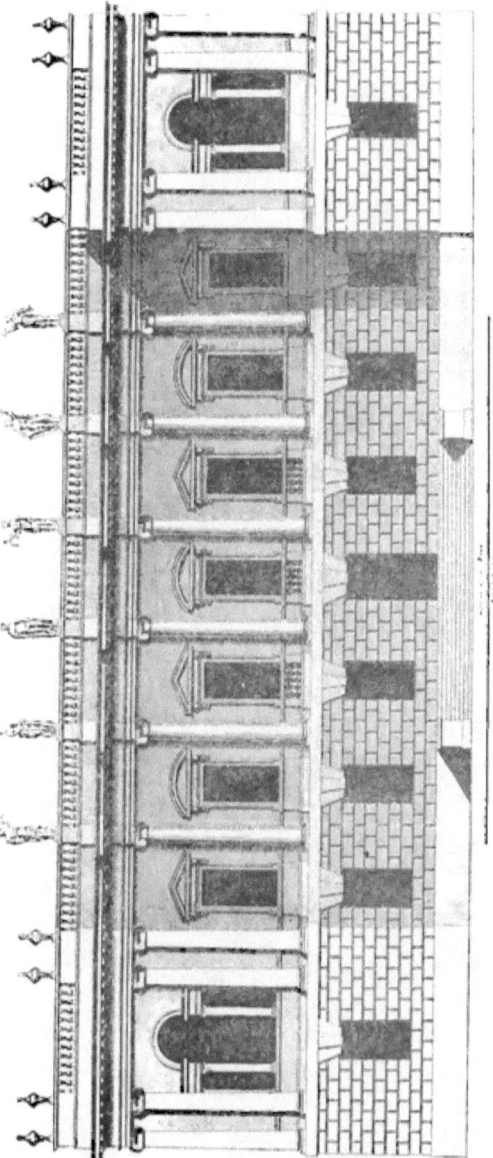

BURLINGTON HOUSE.

the park, lower down the hill. When the Crescent was built the site of the park was a common. It was designed by the younger Wood, and like the Circus consists of thirty houses. A solid basement sustains a series of Ionic engaged columns, those at the ends and in the centre being coupled, the rest single. The effect is good, but owes a great deal to the situation. In any case, it must be acknowledged that neither the Adelphi nor Somerset House has anything like the same stately effect.

Prior Park, a little way to the southward of Bath, is another example of Wood's skill. There are many houses close by in the same style, at Widcombe, for instance, at Corsham, and, in short, wherever the famous Bath oolite can be quarried. But Prior Park is both the most imposing and also the most pleasing. It labours under one serious disadvantage, in facing to the north, and a little neglect covers the noble flight of steps with green mould. The house, which stands on a lofty slope about 400 feet above Bath, consists of a centre and wings, together with outlying offices which are connected with the main building by low walls. The present owners have made some alterations in these arrangements. The portico is very fine, of the Corinthian order, with six columns in front, and two intercolumniations at the sides. The flight of steps is not by Wood, having been added later, but is a very handsome and appropriate feature. The house was built for Ralph Allen, whose name so often occurs in the memoirs of Pope, Fielding, and other great men of the day, and was commissioned with a view of attracting attention to the advantages of Bath stone as a building material. A pretty Ionic bridge in the grounds is copied, perhaps too literally, from one at Wilton, attributed to Inigo Jones.

After the Woods, at Bath, came Baldwin, who designed the

Pump Room, which is not very good, and the Guildhall, built in 1775, which is admirable, but somewhat marred by additions in an anomalous style. These additions are, I believe, condemned, but whether they will be replaced by anything better time alone can show. Architectural taste has died out in Bath.

There was a partial revival of the purest Palladian in London about fifty years ago, by Barry. The only fault of his Reform and Travellers' Clubs in Pall Mall is, that they resemble too closely the Italian buildings from which, professedly, they are imitated. The Reform Club, finished in 1840, is an exact transcript of part of the Farnese Palace at Rome, with the result that the windows, intended for the sunny Italian climate, look too small and dark in Pall Mall. We also miss the beautiful arcaded "loggia" of the original. On the other hand, Barry's proportions are admirable, and he was not content to take a portion of the great Roman house without modifying the dimensions in a similar ratio. In short the design, as adapted, is one of the most perfect examples in London, depending, as it does, not on ornament but on proportion for its effect. The Travellers' has been disguised in stucco and painted, the effect being ruined. It was an adaptation from the Pandolfini Palace at Florence, which is often attributed to Raphael. The critics were wild with delight when this house was first built. After the ill-treatment it has received, we might pass it by daily without remarking its merits, which are undoubtedly very great. It is a question whether the south front is not the better of the two. In it the lower storey is rusticated and the upper windows have balconies. The cornices on both fronts are bold, and were delicately ornamented, but the stucco spoils all. There is no balustrade. Barry was so pleased with this

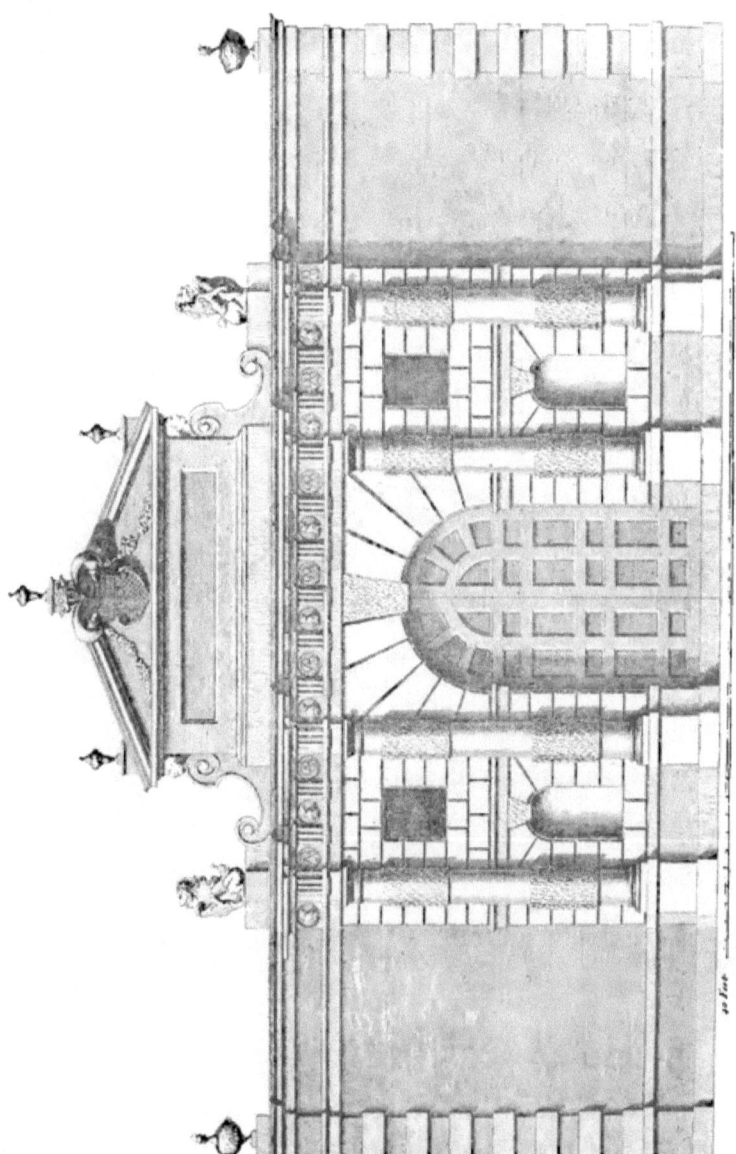

GATE, BURLINGTON HOUSE.

little building that he published a volume about it, with views and drawings by Hewett. Bridgewater House, also by Barry, is much less ornate than the clubs, but on the whole it is greatly to be admired, and shows well beside Spencer House in the Green Park. It seems a pity that Barry's later years were taken up with a building in a style which he did not understand; but the palace of Westminster attests nevertheless the greatness of his powers, even if it makes us regret the more the purpose to which they were put. It has often been remarked that we employed our most eminent Palladian to build in Gothic, and next employed our greatest Gothic architect to build in Palladian; but the Foreign Office in Whitehall, a grotesque structure, shows us that Scott had none of the adaptability or versatility of Barry. A fine club in Pall Mall—the Army and Navy—is by Parnell and Smith, and is copied from another design of Sansovino, the Cornaro Palace at Venice. Both it and the Carlton, by Smirke, in imitation of Sansovino's Library, would look better without balustrades, which go far to spoil the beautiful cornices. Very few other buildings in this style worthy of notice have been erected in London since the time of Barry. I do not wish here to speak of contemporary architecture more than is absolutely necessary, and need not therefore repeat my opinion of the recent buildings in the City or in some of the west end streets. Costly materials, it is proved over and over again, will not make handsome buildings unless the architect knows his art. In the provinces things have, on the whole, been better. St. George's Hall set a fashion in Liverpool which has been followed in the public library and the picture gallery near it, as well as in some institutions of less note. The style flourished for a time at Newcastle, where Dobson designed the admirable railway station and several streets of remarkable beauty and stateliness. Manchester has been unlucky; and except the cathedral and the exchange, contains nothing one would willingly look at twice. The Town Hall must be characterised as frightful, and the Law Courts are only a shade better. Edinburgh never greatly affected Palladian, though it contains some good examples of the genius of Adam; and Glasgow, though it has produced some good architects, has let them practise elsewhere. One of the most original and brilliant professors of the Grecian style was Thomson, of Glasgow, but very few of his works are extant. Belfast also is wanting in good architecture; but before the end of the eighteenth century Palladian art already flourished where we should hardly have expected to find it. Dublin took up the tradition let drop by Bath. A hundred years have elapsed, and we have pleasing evidence that it flourishes there still. The charming group of new buildings in Kildare Street would be more encouraging if we had not to contrast it with the public buildings of the same time and kind in London, and especially at South Kensington. The earliest and certainly the best known Palladian building in Dublin is the Bank of Ireland, originally erected for the Irish Parliament. The architect seems to have been Sir Edward Lovet Pearce, the Surveyor-General, and he deserves credit for a piece of originality not to be matched in the three Kingdoms. The plan is oval, or nearly so, the wall being plain and without openings except on the southern face. There is no basement, but a row of engaged Ionic pillars rises directly from a low plinth. The southern face is very curious. A rectangular opening discloses a kind of courtyard, round which the order is

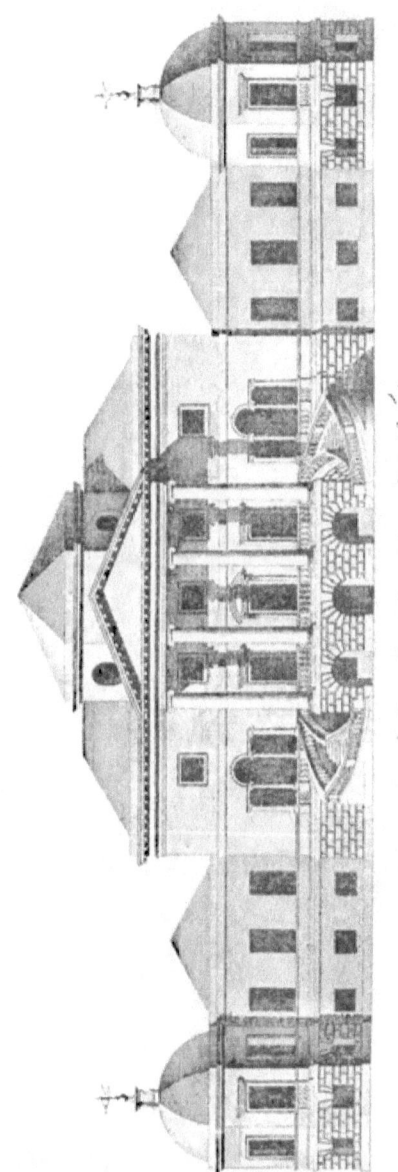

WROTHAM, MIDDLESEX. BY ISAAC WARE.

continued by an open colonnade, a handsome portico and pediment terminating the view. Two smaller pediments, with archways, are at the entrance to each colonnade. The writer of Murray's *Handbook* thus describes this part of the building:—
"It consists of a magnificent Ionic front and colonnades, the centre occupying three sides of a receding square. The principal porch is supported by four Ionic pillars, and is surmounted by a pediment with the royal arms and a statue of Hibernia, with Fidelity and Commerce on each side, the last two having been modelled by Flaxman. The open colonnade extends round the square to the wings and is flanked on each side by a lofty entrance arch." It appears that this grand colonnade and entrance front was the first portion erected. A second portico is on the eastern side, and was built by James Gandon, one of the editors of the fourth and fifth volumes of the *Vitruvius Britannicus*. It looks strange, for it is of the Corinthian order, as desired by the Lords, to whose chamber it gave access. Gandon is said to have described it contemptuously as "the order of the House of Lords." It has by no means a bad effect, being another example of the axiom that incongruous objects, if they are good equally, tend to picturesqueness. Gandon carried out the "order" in 1785, and two years later Parke built a handsome Ionic portico for the House of Commons at the western side. Gandon completed another important design in Dublin. This was the Four Courts, commenced in 1776 by Cooley, who had designed the Dublin Royal Exchange. Cooley died when he had only begun the work, and Gandon finished it in 1800. It is undoubtedly a very fine edifice, and like the Bank shows great originality of plan. An experiment has been tried at Melbourne in Australia by which a similar dome and portico are placed on a conspicuous

hill, and we can judge how much the Dublin courts lose by their situation in a hollow, close to the river, which is here only a noisome sewer. The plan is described by Dr. Walsh, the author of a *History of Dublin*, as one which may be "distinctly delineated in the imagination by figuring a circle of 64 feet diameter, inscribed in the centre of a square of 140 feet, with the Four Courts radiating from the circle to the angles of the square." The columns round the hall are of the Corinthian order, and so also are the columns on the exterior supporting the low dome, and those of the portico.

A little farther down the same quay is one of the largest and most costly of the Palladian buildings of Dublin. The Custom-House was designed by Gandon, and, as it stands by itself, has four fronts, of which that to the south is the best. It has a fine Doric portico, on either side of which is a lofty arcade occupying the basement. From the centre of the building, which is 375 feet long by 205 in depth, rises a cupola, evidently constructed in imitation of the two which Wren placed over the chapel and hall at Greenwich, and rising to a height of 113 feet.

Trinity College is another extremely satisfactory building. The front to College Green is plain but in good proportion. Adjoining the front to the south is the Provost's house, designed after Burlington's house for Marshal Wade. The first college quadrangle, with the Examination Hall and the Chapel, has an irregular but most satisfactory effect. The design was sent over by Chambers, and was carried out by Mayers. The Chapel is by Cassels. On the south side is the Library, formerly a magnificent example of the style. Some forty years ago fears were expressed as to the stability of the flat roof, over 200 feet long, and a stupid expedient was adopted for supporting it,

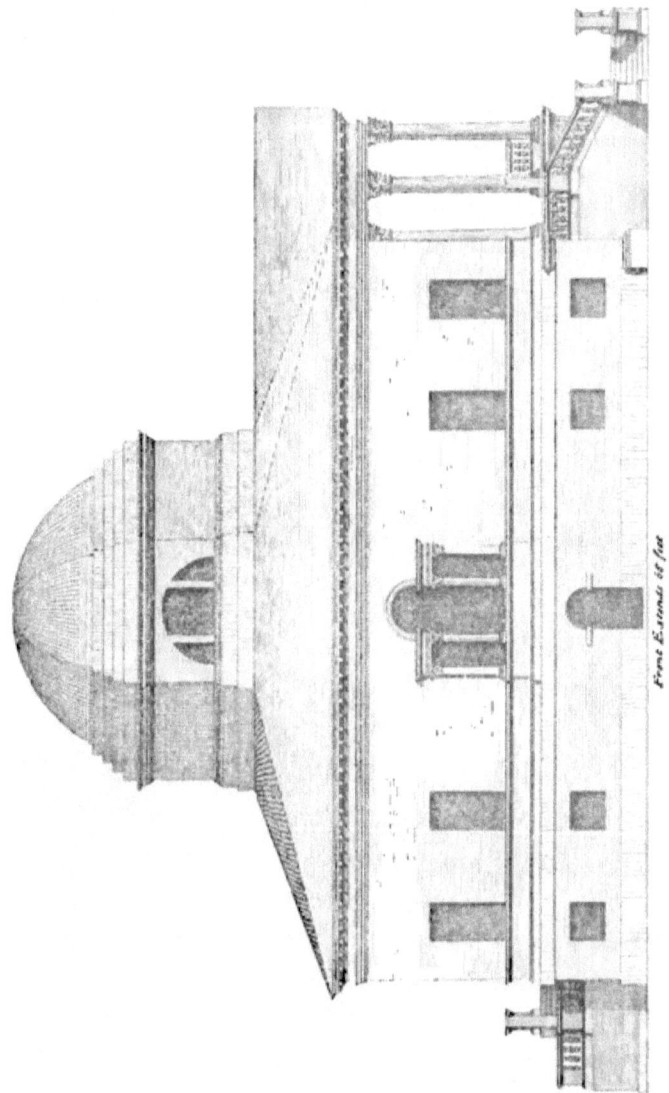

Front Extends 68 feet

VILLA, CHISWICK. BY LORD BURLINGTON.

by which the beauty of the whole building is marred. There used to be a long arcade below, as a cloister, but the arches are now filled. The Library was greatly altered and a new roof made in 1860. In the centre of the quadrangle is a very graceful campanile, by Lanyon, built in 1854, as if to show that the Palladian tradition survived till then. It survives still, as I have remarked, and unquestionably the best contemporary example in the three Kingdoms is the new museum building in Kildare Street. Some people perhaps may assert that the best contemporary examples are to be sought for at the Antipodes. Certainly we have nothing in London to compare with the Post-Office at Melbourne, designed it is said by an amateur. It follows closely the motives of Inigo Jones in his Whitehall drawings, and if it was better situated, for it is in the bottom of a valley, would deserve and obtain a world-wide reputation.

It is necessary, in conclusion, to make some allusion to the style which succeeded or superseded that of Chambers and Gandon and Barry. Gandon and Chambers flourished before, and Barry after, the issue of Stuart and Revett's great work on the architectural remains of Greece, and, in particular, of Athens. "Athenian" Stuart, as he was called, must be regarded as the originator of a taste for Grecian architecture. He did not succeed with it himself, nor did the taste for it last. The picturesqueness of which Thomson of Glasgow has shown it was capable was wholly missed, and some designs of Hardwick, for the railway station at Euston Square, are almost the only good work of the kind which was produced in London. Hardwick was also, it may be remembered, the designer of one of the very few good Gothic buildings of the revival, a library at Lincoln's Inn, the beauty of which was after his day much marred by some additions in a style precisely similar, but without the proportion which in Hardwick's mind was essential to good architecture whether Greek or Gothic. St. Pancras Church became a warning—a sign-post: "No thoroughfare this way." The architects—the Inwoods, father and son—had imbibed to the full the pernicious doctrine that absolute symmetry was necessary, especially in a public building. The details of the design are Ionic, and the whole church, which was finished in 1822, was supposed to be a composition from the Erechtheum and the Tower of the Winds at Athens. Like the Erechtheum, it has a hexastyle portico. Exactly matching at each side are the lateral Caryatid porticoes. In the original, I may say, in accordance with the strictest Greek taste, the two porticoes are different: a charming feature in itself. Here, at St. Pancras, the design of one side is exactly repeated at the other. "Four statues of females of colossal size stand on a continued plinth, in the middle of which are folding doors of iron, closing the entrances to the vaults or catacombs beneath the church." The result is tame to the last degree. So afraid were the architects of admitting the slightest trace of freedom, that each of the Caryatides has a water-jug in one hand and an inverted torch in the other. They are not even carved in marble or stone, but are cast in terra-cotta. In addition to the deplorable effect thus produced, the tower was designed and arranged so as to spoil the one good feature, namely the portico. It rises to a height of 157 feet, and is constructed of a series of temples piled one on the top of the other, each octagonal and peripteral, and generally modelled after the Tower of the Winds. On the top of all "in lieu of the Triton and wand in the original, the symbolical representation of the wind, which terminated the composition, a

SECTION, CHISWICK. BY BURLINGTON.

2 M

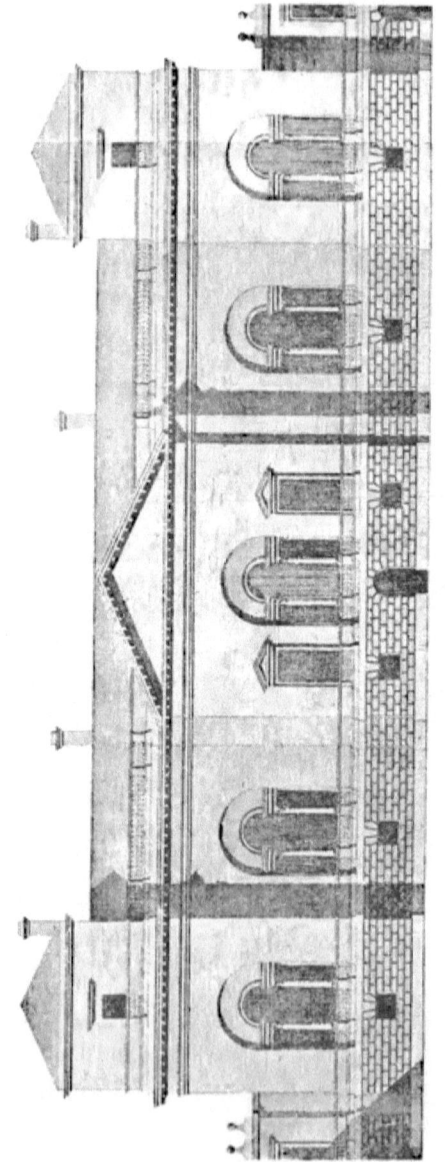

HOLKHAM. NORFOLK. BY KENT.

cross, the great emblem of Christian worship, is placed." Not only is this ridiculous tower very ugly in itself, the proportions of any one storey of it by no means setting off the proportions of the next, but it is put right in the middle of the portico, which is completely crushed by it in every possible view. A church at Glasgow, by the late "Grecian Thomson," shows how such a feature as a tower can be managed in this style, and also where it should be placed. St. Pancras shows us how it cannot be managed and where it should not be placed. So careful were the Inwoods of the details that one of them proceeded to Athens and obtained leave to make casts of parts of the Erechtheum, and, perceiving that some of the columns had been made of the green marble called *verd antique*, he had them imitated in *scagliola*, that is, stucco, in the new church. At first the new design was seriously praised. Anything that professed to be and apparently was pure Grecian, exactly and accurately copied in every part from the best originals, must be good, and not only was St. Pancras admired by the professional critics, but even by some people of taste. But the hour of triumph was very transient. It was soon recognised that the best details will not make up a good building, unless there is a great deal besides in it. There could be no question that St. Pancras church was, from every point of view, a failure,—almost a ludicrous failure. A fresh attempt was nevertheless made by Smirke at the British Museum. "Nothing," says Fergusson, "can well be more absurd than forty-four useless columns, following the sinuosities of a modern façade, and finishing round the corner." There is a somewhat better portico at the Fitzwilliam Museum at Cambridge, by Basevi; but it is better because there is no make-believe, and because it is only one feature of a building and not the building itself. St. George's Hall, at Liverpool, is the most successful example of this school, and it is not so much Grecian as Roman. It is by Elmes, and is an adaptation of one of the Thermæ of imperial Rome. "The principal façade," says Fergusson, "is ornamented by a portico of sixteen Corinthian columns, each 46 feet in height, beyond which on each side is a 'crypto-porticus' of five square pillars, filled up to one-third of their height by screens, the whole being of the purest and most exquisite Grecian, rather than Roman detail." The south front has a handsome octastyle portico. The north front has an apse, and the west is studiously plain, but with Grecian mouldings and other suitable features. It will be seen that Elmes did not share the Inwoods' opinions on symmetry, and this one building would be sufficient to show of how little value it is to the architect who knows what he is about. Before St. George's Hall was completed, in 1854, Grecian architecture may be said to have died out in England. The church of St. Pancras inflicted a heavy blow upon it. The façade of the British Museum is a little better, but it is impossible not to agree with Weale when he wrote in 1853, "The imitations of the most sublimely beautiful productions human art has ever achieved, or is likely to achieve, are now shunned by all for their intense ugliness." The fact is, we had the style and we had the demand, but, and this is after all the important thing, we had not an artist to answer to the call. True, as I have already remarked, an architect in Grecian of great power, named Thomson, showed in some admirably-proportioned designs what might have been done, but very few of his learned and delicate designs were carried out before his death, some five-

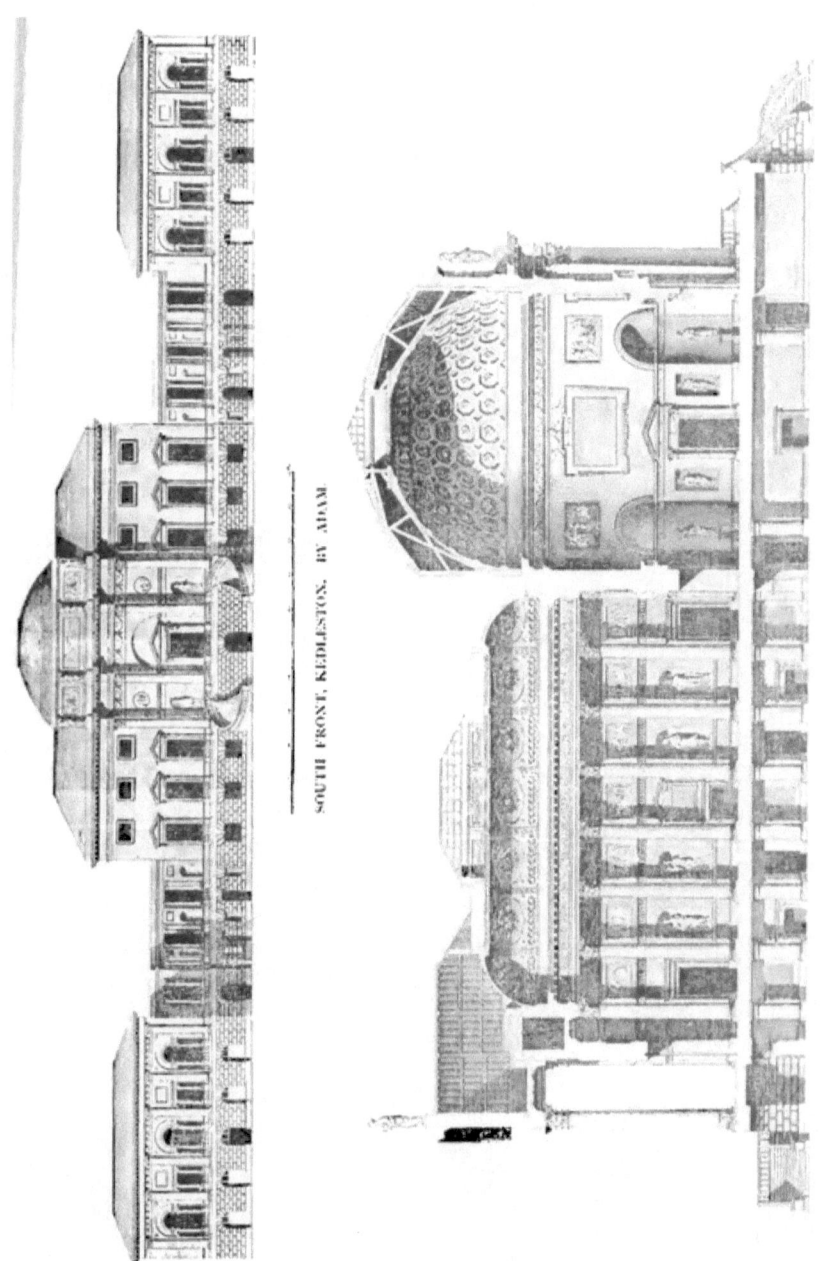

SOUTH FRONT, KEDLESTON. BY ADAM

PRIOR PARK, BATH.

and-twenty years ago. He seems to have cared less for symmetry than for proportion, and was particularly noted for his management of blank walls. But in London it was not altogether the want of a man. Decimus Burton, who designed the screen at Hyde Park Corner, was very fair. His second design for a corresponding screen to be placed at the end of Constitution Hill is even more beautiful, but was never carried out, although it would have been in every way preferable to the present stupid arrangement. The arch, recently removed and rebuilt, in a wholly unsuitable situation, on a slope, standing cornerwise to the roadway, was also his, and though good, is somewhat heavy. There are also several of his designs in *Public Buildings of London*, such as "Mr. Greenough's Villa," the Colosseum, terraces in Regent's Park, and so on; but all were executed in stucco and look well only on paper. He acted as architect to Kew Gardens, where he was very successful with the great palm-house. Nash, also, had ideas. We see his work in Regent Street; but his genius, if he had any, was smothered in lath and plaster. He was much more Palladian than Greek; but if his masterly Quadrant had been carried out in stone, it would have equalled anything of its kind in Europe. Robert Abraham built the County Fire Office in Regent Street, an adaptation from Inigo Jones. Many of the buildings in this neighbourhood as far up as the Regent's Park are very Grecian in design; but what killed Grecian as a modern and feasible style was stucco. If we open any architectural book of 1825 or about that time, we read of grand buildings in Pall Mall, Regent Street, and Regent's Park. In Shepherd's *London in the Nineteenth Century* they are seriously described and favourably criticised by Elmes; while the views,

especially of Regent Street and of Cumberland Place, are,
literally, most imposing, that is, if you do not examine them too
closely : for all these gorgeous palaces fail in the details—as
indeed they must. They are not real. They stand to architecture as scene-painting stands to landscape. They are contrived
to produce an effect, and they produce it until you come near
enough to see that they are constructed throughout of Portland
cement and the like, and that the capitals and mouldings are
cast. Yet, on paper, in views like Shepherd's and descriptions
like Elmes's, they appear splendid. I remember seeing in a local
guide-book—it was in Australia—an account of a Wesleyan
church in a town not too far off for a visit by rail. The building was described as being "in the Grecian Doric style. . . .
The entablature is divided into architrave, frieze, and entablature
cornice, the frieze being enriched with triglyphs and drops, the
former capped. Under the corona of the cornice is a handsome
modillion band, with mutuals central with each pilaster. The
centre portion of the building, by projecting pilasters about a
foot, enables a feature to be made in the front—the balustrade
terminating with a pediment. The use of smaller pilasters
centrally situated, with cornices and pediment, secures a porch
showing prominently, and an arched recess over gives a tone
generally to the projecting portion." There was a great deal
more and a similarly florid account of the interior. I naturally
journeyed off to see this remarkable example of colonial taste. I
found an ordinary little Dissenters' meeting-house, of stucco,
painted drab. I did not look at it twice, but had time to recognise that the description was perfectly correct.

This kind of thing, I am sure, this elevation of the make-
believe to an equality with the real, helped largely to ruin

REFORM AND CARLTON CLUBS, PALL MALL.

CHURCH AT GLASGOW. BY THOMSON.

the practice of Grecian, and to prepare the minds of people of taste to expect great results from the Gothic revival. There was in future a sweet simplicity in classification. Gothic, of course, was Gothic. But Grecian described everything that was not Gothic — Regent Street first, and after it Whitehall and St. Paul's and Burlington House. No invidious distinctions were made. All were classed in the same category. They were not Gothic.

After many years, however, and much eloquent writing and elaborate designing, people are beginning to discover that there is a style that is neither Grecian nor Gothic, a style, too, which, unlike either, encourages a designer to be original, and desires him to go forward, and not backward, and which, though it is by no means new, does not prohibit novelty. The Palladian style is about 400 years old, yet its admirers are not obliged to build after a 400-years-old pattern. They are able to use the building appliances of the day. Without any falsification or make-believe they can employ every new invention for heating and lighting, for ventilation and sanitation. The style often in this country called "Queen Anne" is included in it, and there are other names: but Inigo, Wren, and Burlington were, and acknowledged themselves to be, under the influence of Palladio rather than of Bramante or Vignola, or Sansovino, or San Gallo, or any other great Italian of that time.

The great drawback to the use of this style at the present day lies in the necessity for careful proportion. It comes into competition, not with Grecian, as it was understood when George IV. was king, nor yet with Gothic as practised by Scott and Street, neither of whom seemed to have known what proportion meant, but with the new wilfully ignorant school, the "know nothings" of English architecture; and so, in the closing words of this chapter and this book, I come back to what I began with. If the British public, the employers of the British architect, insist on learning and beauty, whether Gothic or Palladian, or Grecian, it daily becomes more likely that they will get it. But, if they profess not to care, we may go on as at present with cathedrals like Truro, with palaces like Buckingham Palace, and with private houses like Grosvenor Place.

INDEX

Abraham, Robert, 273
Adam, R., 242
Alban, St., 187, 212
Albert Memorial, 6, 24
Alcove at Kensington, 163, 172
Allen, Ralph, 249
All Hallows, Bread Street, 160, 184
All Hallows, Thames Street, 184
All Souls' College, 35, 220
Almerico Palace, 101
Alphage, St., 227
Amesbury, 139
"An arch never sleeps," 19
Anne Boleyn, 45
Anne of Cleves, 45, 80
Anne of Denmark, 46, 115
"Anne, Queen," style, 164, 168, 171, 172, 279
Anomalous style, 12
Antholin, St., 180, 211
Arch Row, 121
Archer, 227
Architecture, modern, 3, 15, 279
Arundel, Thomas Howard, Earl of, 113
Ashburnham House, 110, 135
"Athenian" Stuart, 261
Aubrey, 68
Audley End, 55, 63

Baldwin, 249
Bank, the, 245

Banqueting House, Whitehall, 110, 122, 128
Barry, Sir C, 4, 250
Basevi, 267
Bath, 36, 228, 246, 249
Bayswater Road, houses in, 9
Benet, St., 135, 144
Bernini, 157
Blomfield, Mr. Reginald, quoted, 59, 67, 144
Borley, 83
Bradford-on-Avon, 64
Bray, 15, 23, 24
Bride's, St., 202
British Museum, 268
Browne, Sir Anthony, 83
Brympton, 97, 139
Burges, W., 34, 197
Burghley, 73, 75
Burlington, Earl of, 8, 140, 144, 215, 228
Burton, D., 273

Caius College, 87, 92
Cambridge, 11, 87, 139, 152
Campbell, C., 215, 228, 230
Canova, 204, 209
Canterbury, 20
Carlton Club, 253
Chambers, Sir W., 105, 233, 241
Chancellor, Mr., 83

Charles, Plymouth, 40
Charles I., 127
Charles II., 171, 194, 198
Charlton, 139, 170
Chelsea, 167
Cheshire Houses, 71
Chichester Cathedral, 6
Chiswick, 238
Cibber, C., 172
Clark, Mr. J. W., 88
Cobham, 52, 55, 63, 74, 139
Coleshill, 140
Collcutt, Mr., 12
Cooley, 257
Corsham, Wilts, 39, 249
Covent Garden, 110, 132, 228
Cromwell, Oliver, 151
Cromwell, Thomas, 45

DENHAM, SIR JOHN, 152, 153
Devonshire collection, 109, 116, 144
Dormer Monuments, 83
Dublin, 254, 261
Duke's House, the, 64, 68
Dunstan's, St., 211
Durham, 20

ELIZABETHAN HOUSES, 54
Elmes, 204, 268, 273
Ely, Bishop of, 36, 151, 152, 196
Ely, chapel of Bishop West, 46
Ely, monuments at, 83

FELL, DEAN, 32
Felstead, 83
Fergusson, James, quoted, 125, 135, 157, 207, 219
Fletcher, Professor, quoted, 106
Four Courts, Dublin, 257
Fulmer, 40

GALILEE, Durham, 20
Galleries, Wren's, 201
Gandon, James, 257, 261
Garbett on design, 98
George's, St., Hall, 268
Gibbons, G., 120, 210
Gibbs, James, 219, 227
Glasgow, 254, 267
Gloucester, 22
Gotch, Mr., on Elizabethan Buildings, 52, 74, 75, 116
Great Chalfield, 36, 64
Great St. Mary's, 35
Greenwich, 139, 152, 158, 167, 170
Gregory, St., 131
Gresham's Exchange, 87

HADDON, 55, 59
Halliday, 39
Hampton Court, 28, 33, 43, 96, 163, 171
Hardwick, 97, 261
Harewood, 241
Haveus, 87, 92
Hawksmoor, 28, 35, 171, 215, 219
Haydocke, 93
Henry VII., 3, 21, 79
Henry VIII., 31, 72
Heriot's Hospital, 46, 115
Holbein, 45
Holkham, 233
Holland House, Gate of, 140
Holt, Thomas, of York, 31
Honour, Gate of, 91
Hungerford Hospital, 39
Hyde Park Corner, 273

INWOOD, 262

JAMES, JOHN, 227
James I., 122, 126, 127

Index

James's, St., 45
Jewitt, Mr. O., quoted, 31
John's, St., Oxford, 32
Jones, Inigo, at Oxford, 32; in London, 47; birth, 113; career, 116, 147; death, 143

KATHARINE CREE, ST., 47, 131, 187
Kenilworth, 72
Kennington and Kensington, 139
Kent, William, 215, 228, 233
Kew Gardens, 242, 273
Kildare Street, Dublin, 254, 261
Kingston House, Bradford, 68
Knole, 53, 59

LANSDOWNE HOUSE, 245
Laud, Archbishop, 47, 187
Law, Mr. E., quoted, 43, 171
Lawrence, St., 203
Lichfield Cathedral, 211
Lincoln, 46
Lincoln's Inn, 241
Lindsey House, 110, 121
Liverpool, 253, 268
Lomazzo, 93
London, modern architecture of, 7, 13, 273
London, Tower of, 20, 44, 54
Longland, Bishop, 46
Longleat, 71
Lothbury Court, 245

MANCHESTER, 12
Marlborough House, 163
Marney, Layer, monuments at, 80
Martin, St., in the Fields, 221
Mary, St., Abbots, 10
  ,, ,, Abchurch, 210
  ,, ,, Aldermary, 48, 179
  ,, ,, Great, 35

Mary, St., Le Strand, 98, 221
  ,, ,, Somerset, 179, 203
Matthew, St., 153
Mildred, St., 209
Monument, 163, 170
Morden College, 170

NEVILE'S COURT, 155
New Law Courts, 5
Newcastle, Duke of, 222
Nottingham House, 139

OPIE, JOHN, 9, 160
Orangery, 173
Orders of Architecture, 100
Ornament out of place, 5
Osyth, St., 83
Oxford, 28, 32, 35, 115, 139, 154
Oxford, Lord and Lady, 222

PADUA, JOHN OF, 68, 73
Palladio, 98, 99
Pancras, St., 262
Paris, Wren in, 156
Parnell and Smith, Messrs., 253
Paul's, St., Cathedral, 187, 200
  ,, Covent Garden, 131
  ,, ,, Old, 131
Pembroke College, 152
Phillimore, Miss, quoted, 192
Plymouth, 40
Prior Park, 249
Proportion, Palladian rules of, 105

QUEEN ANNE style, 164
Queen's Square, Bath, 246
Quincy, Quatremere de, 99, 113

RANGER'S LODGE, 242
Regent's Park, 273

Reredos at Whitehall, 131
Rich Memorial, 83
Roehampton, 242
Rowe, Sir H., 39
Ruskin, Mr., 6

SALISBURY CATHEDRAL, 6
Sansovino, 99, 253
Savoy, Chapel Royal, 45
Scott, Sir G. G., 6, 32
Sedding, the late Mr. John D., 15
Sedding, Mr. Edmund H., 209
Sheldonian Theatre, 154
Shelford, Little, 92
Soane, Sir John, 131, 136
Solomon, Song of, 95
Somerset House, 121, 241
Stephen's, St., Wallbrook, 11, 204, 209
Stevenson, Mr. J. J., quoted, 22
Stewart monuments, 83
Stone, Nicholas, 121, 128, 140
Strong, Thomas and Edward, 164
Stucco, reign of, 273
Swithin's, St., 211

TAYLOR, MR. ANDREW, quoted, 183, 184
Taylor, Sir R., 241, 245
Temple Bar, 170
Temple, Middle, 170
Theobalds, 170
Thorpe, 73, 92, 93
Tom Tower, 33, 159

Torregiano, 44, 79
Tower of London, 20, 44, 54
Trinity College, Cambridge, 155
    ,,   ,, Dublin, 258
    ,,   ,, Oxford, 155
Tuscan style, 105

VANBRUGH, 169, 216, 219
Van Eyck, 94
Vardy, 227
Venice, 102
Vernon, Dorothy, 56
Verona, 105
Vicenza, 102
Vitruvius, 102, 131

WADE, GENERAL, 228
Wadham College, 28
Ware, I., 234
Webb, J., 143
Westminster, 21, 28, 80
Whitehall, 122, 129
Widcombe, 109, 249
Williams, 44
Wolsey, 33
Wood, 246
Wraxall, South, 23, 36, 53, 64, 84
Wren, Sir Christopher, 34, 151, 212, 219

YORK, DUKE OF, 191, 193
Young, John, 44, 79

THE END.

www.ingramcontent.com/pod-product-compliance
Lightning Source LLC
Chambersburg PA
CBHW030819230426
43667CB00008B/1287